30 Minute Risk Assessments – Control Based Risk Analysis (CoBRA)

David G. Sheriff

Copyright © 2016 David Sheriff

All rights reserved.

ISBN:1500131512
ISBN-13:9781500131517

DEDICATION

To my lovely wife Carolyn who tolerates so much and only asks that the cats are fed and there is always a good selection of red wine to choose from.

CONTENTS

	Introduction	1
1	Background	6
2	Shuttle Connection	10
3	Loss of Columbia	14
4	Trouble With Risk Assessment	17
5	Cattle or Fences	22
6	Understanding Our Controls	26
7	Controls and Processes	34
8	Process Design Basics	41
9	Making Some Progress	44
10	All Hail the Process Owners	49

11	Can't Fix Being Human	57
12	Keep It Simple and Quick	62
13	Getting Ahead of Problems	68
14	Changes Begin to Pay Off	78
15	CoBRA Risk Assessments	82
16	Metrics Indicate Success	92
17	Runt the Twelve Item Checklist	100
18	Little Things Can Bite Hard	126
19	No Problems, Think Again	132

| 20 | Accepting High-Risk Situations | 140 |
| 21 | To Sum Things Up | 146 |

INTRODUCTION

I am writing this book with the purpose of capturing what a group of Corrective Action Engineers, of which I was a member, learned during 24 years of investigating the mistakes made by the thousands of workers who prepared the Space Shuttles for launch at the Kennedy Space Center. This book is going to be as short and as simple as I can possibly make it because I believe that the less complicated something is, the more useful it becomes. If what my group learned was difficult to understand and required a large number of pages to try to explain, I would not even bother to waste your time sharing it with you.

This book was not intended for people who study human error. In fact, this book should be considered to be more of a "shop manual" for those individuals whose job requires them to actually deal hands on with human error on a daily basis.

This book was written for managers that want to make sure their processes will be performed correctly and their people are working safely with no one at risk of being hurt as a result of a human error.

This book was written for people who get handed a human error incident and told to investigate it, find out the real reason for why it happened, and make sure it never happens again.

This book was written for the person whose management tells him that the workforce is making far too many mistakes and they expect him to find out why and do something about it.

This book was written for the engineer who is responsible for some critical process and wants to limit the possibility of a human error occurring while the process is being worked.

It was apparent to my group after thousands of investigations that whenever a large critical process had problems we seemed to always be able to trace back and find numerous contributing process failures involving smaller, less visible, supporting processes. We learned that you cannot depend on a large critical process to be performed problem-free until you can first count on all of your smaller supporting processes to also be problem free. In our situation, these critical processes relied on hundreds, if not thousands, of smaller supporting processes. We had to have a way to find and fix the flaws and weaknesses in the smaller, less visible, supporting processes before we could have complete confidence in our larger, more critical, processes.

It was obvious to us that the very best tool for finding problems in any process is a risk assessment, but the risk assessment methods being used by our company were very time-consuming and they were not something every employee was capable of doing. We also encountered a more serious problem with these risk assessments in the fact that we could not get separate groups, even when each group was given the same exact information, to arrive at the same final risk values. You can't feel a risk assessment is reliable when one group finds the risk value to be high and another group using the same information finds the risk value to be low.

The only way any company can get the thousands of necessary risk assessments performed on all processes is to have the local workgroups that actually work these processes, the true experts on these processes, perform the risk assessments themselves. Since these local workgroups still have their regular functions to perform, you can only ask them to devote a minimal amount of time to doing risk assessments. In order for local workgroups to perform their own risk assessments, the assessments had to be simple, accurate, and quick to do. The 30 Minute Risk Assessment concept came about just because it meets the basic requirements of being simple, accurate, and quick to perform.

What my group learned applies to a workforce of 2 or even 20,000. Any company with a human workforce has to be concerned with workforce error. This book explains how the 30 Minute Risk Assessment concept came about, the basis on why it works, and how any organization can use it to assure all of their processes can be counted upon when needed.

The group of engineers I was part of did not get involved in hardware or hardware design failures. We focused almost entirely on workforce human error. Our group realized early on that there was a great deal of information available on hardware reliability, but we could find little usable information on human reliability. What my group could not find was something out there to clearly and simply tell us exactly what we had to do on a day to day basis to eliminate or reduce the incidents of human error we were seeing. To us, it was like a car mechanic just trying to find out how to change a water pump and all of the information available was on the thermodynamic principles involved in the internal combustion engine. It was not that any of the information out there was incorrect, it just didn't help us for what we needed. Since we could not find anything to tell us what we specifically needed to do to address our human error problems, we had to come up with it on our own.

Over the years, we developed simple concepts and methods to assist us in performing human error investigations, risk assessments, process design, and also with the determination of suitable corrective/preventative actions. Our group had to develop concepts and methods that were quick and easy to use since we dealt with a workforce of several thousand and our NASA counterparts expected some level of preliminary investigation and understanding of every workforce error within hours of the occurrence. NASA also expected the fixes for any human error issue to be implemented in a timely manner and be effective.

I am completely confident that anyone having read this book will look at workforce error differently. A reader will be able to recognize those processes where errors are likely to occur that they would not have identified previously. A reader will understand how risk assessments can be performed that are quick, simple, repeatable, and capable of identifying high-risk situations correctly. These risk assessments will also be suitable for clearly identifying exactly what actions need to be taken to reduce the risk identified and assist in prioritizing the order for addressing the required actions.

I understand that there are a lot of varying opinions related to how human error should be considered and the best way of doing risk assessments. The purpose of this book is not to say one way is right and another way is wrong. I honestly believe the right way is whatever works best for your particular situation. The advantage of doing the 30 Minute Risk Assessment is that it is fast and easy to do and therefore suitable for use by local workgroups for assessing all of their own processes. Every workgroup should always be evaluating their own processes and identifying and correcting any potential for human error and this book will provide a simple, quick, and accurate means of doing that.

1 - BACKGROUND

I joined the Shuttle Program at the Kennedy Space Center (KSC) back in early 1987. It was a very difficult time back then for everyone involved in the Space Shuttle Program. The Program was trying desperately to recover from the Challenger disaster. The fault of the disaster was determined to be a design flaw with the solid rocket motor o-rings and not the result of any human error by the KSC workforce who processed Challenger for flight. These workers took the loss of the vehicle and its crew in a deeply personal manner. As a result, the workforce that I encountered was without a doubt one of the most highly motivated and dedicated group of workers you could possibly imagine.

I am not sure most people realized that the actual hands-on processing of the Space Shuttle at KSC was not done by NASA personnel. The thousands of workers at KSC that prepared each Shuttle for launch actually worked for private companies under contract to NASA. There were a large number of NASA personnel at KSC, but their main function was oversight of these private companies' contract compliance.

I was hired by the company that had the main Shuttle processing contract. The workforce my company provided numbered in the thousands and they performed the majority of the day to day processing work done in preparing Shuttles for launch. My particular function was being a member of a newly formed group of ten engineers that established the

Corrective Action Engineering (CAE) group. We were formed as a result of findings from the Challenger Mishap Investigation Board that found a need for more emphasis placed on identifying and preventing potential problems relating to the workforce.

NASA at KSC was a terrific group of people to work with. The safety of the flight crews and workforce was without a doubt the single most important issue to NASA and they made that point very clear on a daily basis to everyone involved in Shuttle processing. NASA provided a great overall working environment but they demanded strict compliance to the processing requirements they had established. NASA also continually set new goals and they expected ongoing improvement in all areas of processing.

All problems that occurred or discovered during Shuttle processing were documented by the workforce. The most common issues were things such as flight damage, hardware failures, design problems, equipment wear-out, and workforce human error. These documented problems were known as a Problem Report (PR). These PRs stayed open until the issue was fixed with all actions taken being documented. Many PRs were very complicated and required several hundred pages of actions to reach closure.

My group, CAE, would meet first thing every morning with our NASA Quality Engineering counterparts and review all PRs initiated during the previous 24 hours. We broke all PRs into one of two categories, Preventable and Non-Preventable. The type of PRs that we focused on was the human error type PRs and they all fell into the Preventable category. Since our company provided the workforce, that is where management wanted our efforts to be directed. Hardware failures and design problems were handled by other NASA contracts with other companies. Hardware failures and design problems were PRs that typically fell under the Non-Preventable category and were not addressed by our group.

During these morning meetings, our group would have an initial discussion on each Preventable category PR. One of us would be assigned that PR to be investigated. Within a few hours, a quick preliminary evaluation of the problem was put together to provide to our management or NASA. Actions required to prevent a recurrence of that problem were discussed within our group and suggestions offered. Corrective actions would be agreed upon and the particular engineer working this problem would see to it that the corrective actions got implemented. When the investigation was complete, maybe weeks later, the final findings of the investigation and all actions taken would be entered into a database our group maintained.

This was the basic process that CAE performed from 1987 until 2011 when the Shuttle Program was ended. The group worked well together and there was little turnover of personnel. When they did leave, it was usually to retire. I stayed on and was a member for the entire 24 years that the group existed.

The group developed a wealth of knowledge and

experience on people, the mistakes they will make, and how best to deal with the potential for these mistakes. During these years, thousands of human error incidents were investigated. Causes for these errors were identified and corrective/preventative actions put in place. NASA funded over a half million engineering man-hours for the work our group performed and that does not include administrative support personnel, computers, office supplies, and vehicles.

My group was not created for the specific purpose of gathering data analyzing human error, but for each investigation we performed, we did document into our database what the investigation found and what we did to address the problem. In some ways, what our group did could have been thought of as a 24-year long study of human error issues encountered by a real-life workforce comprised of thousands of individuals that cost tens of millions of dollars to conduct. When you think of it in those terms, it kind of raises the obvious question "What did you learn?" The simple answer to that question is what I am going to relay in this book.

2 - SHUTTLE CONNECTION

The company I worked for provided the workforce that processed the Space Shuttle at KSC preparing it for the next mission after each landing. While this was done per contract with NASA, my company was definitely not a government agency. My company was motivated by the goal of making a profit. In order to keep the contract and maximize the amount of money received, it was important to keep NASA satisfied with the performance of the workforce that was being provided by my company. NASA was very upset when workforce error damaged hardware, delayed schedule, or injured personnel. The potential for financial penalties or loss of the contract always existed and it was very important to my company's management to make sure NASA felt all problems were being adequately addressed.

From the very beginning, the Corrective Actions Engineering (CAE) group realized that we were very fortunate in several areas. We had a workforce that was highly motivated to do things correctly. The astronauts constantly visited KSC to train and interact with the workforce. The workforce understood that the lives of the astronauts clearly depended on the work they performed.

We were also fortunate that the workforce was well compensated for the job they did. NASA wanted to make sure that the Shuttle fleet was being handled by the best personnel available and they assured that funding was there to hire and keep good personnel. The workforce understood they were fortunate to have these jobs and did not wish to do anything that might put their employment in jeopardy.

Regarding work environment, tooling, and schedule pressure, NASA always made sure that the workforce lacked for nothing. The vast majority of Shuttle processing was done inside buildings where the temperature was controlled. To keep humidity low, the air conditioning was always maintained in a range that was pleasantly cool. For the most part, shuttle processing took place in a work environment was very comfortable, with the exception being the personnel that worked on the launch pads who often worked in exposed and difficult conditions.

The workforce was also provided with the highest quality and best-maintained tools available. All measuring devices such as voltmeters, pressure gauges, or torque wrenches, were part of a calibration program that recalled them at carefully determined intervals to maintain and recalibrate them to assure accuracy when used.

Launch schedules were critical to the Shuttle program. From the minute a Shuttle landed from one mission, it was already on a strict schedule to assure it was ready for the next required launch window. Workers were always under schedule pressure, but they clearly understood that they could call a "time out" at any time they felt safety was being compromised.

One policy in particular that NASA established that I believe showed great forethought, was the policy that a worker would not lose their job over making a mistake that they self-reported and did not try to hide. On the other hand, the failure to report something that you knew you did wrong resulted in the strongest of disciplinary actions.

When working on the very expensive flight hardware on the Shuttle it was extremely easy to make a million dollar mistake, but I never saw anyone fired who reported on themselves regardless of the cost of the error. The important part of this policy was that it resulted in numerous instances of hardware damage being reported quickly by the person making the error. This was very helpful to us since in many cases we might have never known the details of how the error happened if the person had not reported on themselves. Besides enhancing flight safety, this policy also allowed our corrective action group to get the accurate story of how the mistake was made that caused the damage.

With so many things in our favor, the Corrective Action Engineering (CAE) group should have had an easy time at eliminating workforce error. Unfortunately, there were some very important elements of Shuttle processing that were working against us. No two Shuttles in the fleet were exactly the same; they all varied from each other in different ways. Also, no two Shuttles ever returned from flight with exactly the same types and amounts of flight damage. In addition, no two Shuttle missions were exactly the same, so each Shuttle had to be configured differently based on the next mission it was to fly.

Shuttle processing was definitely not an assembly line type of operation with consistent and repetitive processes that could be fine-tuned each time they were performed. In Shuttle processing, even the same tasks that have to be performed each time a Shuttle is processed will always be somewhat different than the previous time the task was performed. This lack of task and hardware consistency made workforce error more likely. It also made it more difficult to fix our human error issues when compared to a consistent and repetitive manufacturing assembly line type of environment.

3 - LOSS OF COLUMBIA

The loss of Columbia was devastating to me and everyone else at KSC. It still makes me feel sick inside to think about it, so I am not going to say much. Like Challenger, the Columbia disaster was a design issue and not the result of any error or mistake by the KSC workforce. Not being at fault did not make anyone in the KSC Shuttle processing workforce feel any better about losing Columbia and its crew.

As a result of the Columbia disaster, the KSC NASA management that we dealt directly with had a serious concern. They were worried that in some of our more critical processes, even though we were presently not experiencing serious human error problems, that maybe we were just being "lucky rather than good". NASA KSC management wanted to have much more confidence that the risk associated with workforce error was understood and being adequately controlled in our most critical processes.

Also as a result of Columbia, NASA management at KSC implemented numerous changes. For our group, these changes involved us being told that it was no longer acceptable for our group to just investigate problems after they occurred. NASA wanted a new emphasis on identifying potential problems and taking actions to prevent them before they could occur. NASA understood that a good preventative maintenance program can significantly reduce the likelihood of hardware failures. What our NASA counterparts were now stressing to us was that they wanted something like a hardware preventative maintenance program, but one that would be applicable to our workforce that could identify and correct potential human error situations.

Unfortunately, the preventative maintenance methods that work so well for preventing hardware problems, don't work when dealing with workforce human error issues. If I wanted to know when to expect a valve to fail I could take several samples of that exact valve and under controlled conditions test them to failure and determine the number of cycles the valves could be trusted. But how do you tell when a human is going to fail? Each member of any workforce is totally different. In our situation, no two tasks being performed were ever exactly the same either. How do you get accurate reliability data on humans when no two individuals are alike and no two tasks being worked are the same?

Our NASA counterparts expected us to do something like preventative maintenance on our human workforce, so we had to figure out how to recognize processes with a high likelihood of failure due to human error even if these processes were presently not experiencing any problems. Our NASA counterparts also wanted us to develop metrics that would show that progress was being made in reducing workforce human error as a result of the new preventative actions we would take.

Our group had always investigated incidents after they occurred and then put in place corrective actions to prevent this exact same incident from happening again as well as any similar situations. Over the years, this is how our group had been addressing NASA's requirement for our group to be doing corrective as well as preventative actions. It was now clear that our NASA counterparts expected a lot more from us and we were at a loss as to how to give our NASA counterparts what they were asking for.

4 - TROUBLE WITH RISK ASSESSMENTS

My company was responsible for the workforce and we had a Quality directorate that included Safety, Reliability, Human Factors, and the group I was a member of called Corrective Action Engineering (CAE). After many years of Shuttle processing and human error investigations, there were no longer any "low hanging fruit" left where the potential for human error problems was obvious. The Shuttle Program was by now a very mature program and easily found problems had already been previously discovered and addressed. The number of problems due to workforce error was significantly lower than it was when the Space Shuttle Program began and to us these existing processes appeared to be about as good as we could make them.

The last thing my management in the CAE group wanted to do was to face our NASA counterparts and tell them we were already doing all we could do, with little opportunity for significant improvements. Our management did not want to say that we didn't have the means or methods available to determine in advance where human error is likely to occur in Shuttle processing which would enable us to take preventative actions.

The corrective and preventive methods our group employed were based on an initial problem occurring first and then taking actions to correct that specific situation and prevent a similar failure from happening elsewhere. If a process did not have a history of human error, we considered it to be safe until a problem actually occurred. No history of workforce error was considered by default to be a safe process. We fixed processes when there was a problem. If it wasn't broke, we didn't see a need to try to fix it.

We understood that this left a potential situation where "the gunpowder factory had never had any incidents and was an extremely safe place to work until the day it exploded killing everyone" type of incident. We knew this was not a very desirable situation and NASA was not going to be satisfied until we came up with new methods and tools that could determine "lucky from good" processes.

Our Corrective Action group now had a significant challenge given to us. We had to come up with methods for distinguishing a safe process from an at-risk process in a way that was not based on any kind of history of previous human error problems in that process. This new method also had to be relatively quick to perform due to the large number of processes involved. It also had to be repeatable, because no one could ever have faith in a method where different groups of individuals using this method would come to different conclusions regarding whether a particular process was safe or at-risk. Finally, it had to be able to generate metrics that would show when improvement was being made or not.

Our engineers attended various classes on understanding risk and learning how to perform different types of risk assessments. Using risk assessments as a way of identifying processes having potential problems seemed like the suitable fix for our problem, but we found these risk assessments also had a serious flaw. The way risk assessments were typically being done required the determination of a Consequence value and a Likelihood value. A numerical final risk assessment value would be determined based on how serious the potential consequence would be for a human error and how likely it is that the human error would even occur and result in the negative event. The risk assessment process would typically assign a numerical value (1-5) for the seriousness of the consequence and a value (1-5) for the likelihood of it actually occurring. If the consequence of the error was very serious, for example, loss of life, then you would assign a Consequence value of 5. If you believed the error and consequence were likely to happen the very next time the task was worked, then it would assign the maximum Likelihood value of 5. This task would then have a final risk value of 5 x 5 = 25 and it would be the very highest final risk value possible. It would also be called a RED risk since it was very common to associate a stoplight color of red, yellow, green to simplify risk descriptions. Different companies chose different ranges for RED, but they were always the upper-risk values. Similarly, a task where the potential human error would cause little or no impact, there would be a Consequence value of 1. If it was extremely unlikely for the error and consequence even happening, then it would receive a Likelihood value of 1 and result in the very lowest final risk value of 1 x 1 = 1. This would be called a GREEN risk along with other risk values in the lower ranges. YELLOW risk scores were assigned to be the middle-risk score ranges.

As you can see, with a risk assessment you need two key bits of data. You need the Consequence value of the potential failure and the Likelihood value of the potential failure. With a hardware risk assessment, the Consequence is easy to obtain from the design of the system and the Likelihood is known from previously performed failure testing history of the hardware.

When looking at potential hardware failures, these typical risk assessments work well, but they are very lacking when it came to trying to determine when a human error was going to occur. For potential hardware failures, these typical risk assessments are a very repeatable, accurate, and useful tool. When dealing with workforce human error, they have a very serious flaw. While the Consequence value could be accurately determined by a system design evaluation, the required value for Likelihood is usually only based on some type of intuitive best guess. Having to make a best guess, even an experience-based best guess, in order to get your Likelihood value was, in my group's opinion, an unacceptable fatal flaw. We were having to guess at Likelihood values such as 1 in 1000, or 1 in 5000, or maybe 1 in 50000. Or we had to distinguish if the occurrence was likely, unlikely, somewhat unlikely, very unlikely, or highly unlikely to happen. Likelihood values based on a guess prevent these types of risk assessments from being repeatable, accurate, or even very useful.

We still saw the risk assessment concept as

potentially the best solution for what we needed, but we realized that doing a risk assessment for workforce error had to be considerably different than a hardware risk assessment. Our group agreed the key to a useful risk assessment would be a Likelihood value that was accurate, repeatable, and easy to arrive at, but we had no idea on how to get there.

5 - CATTLE OR FENCES

Our group needed to come up with new ideas in order to arrive at risk assessment Likelihood values that were accurate, repeatable, and easy to determine. We struggled with this problem for months. We could not get past the fact that we already had data on thousands of human error incidents and one thing was obvious, humans are very complicated and they all have the potential of making a mistake at any time. Whatever we came up with was going to have to be based on the realization that humans are always going to be capable of making a mistake at any time, so we needed to accept and deal with that potential.

Shuttle processing involved many tens of thousands of separate tasks that were required for a successful and safe mission. Some of these tasks were very complicated and required extensive decision-making skills on the part of the workforce in order to be done correctly. We were very fortunate though because we had the kind of intelligent and highly motivated workforce that most companies can only dream about ever having. Still, human error was a recurring problem.

One very useful observation from our data was that even though human error was common, it was rare for human error failures to occur in our more critical tasks. On the few occasions when a human error did occur in a critical process, it was always discovered in time to avoid a serious consequence. The limited number of errors that did occur in critical processes rarely resulted in injury or serious damage to critical hardware.

The fact that human error was occurring in any of our processes, critical or non-critical, led NASA KSC management to their concern that we had been lucky up until now and it was only a matter of time until we had a human error in one of the critical processes with catastrophic results. An initial assessment of this data might lead to a conclusion that the workforce was just more careful when working critical processes and less careful working the non-critical processes, but this was not the case at all. This was a unique workforce that didn't look at some processes as needing a high level of care with other processes warranting a lower level of care. These workers took a great deal of pride in every action they performed. To their way of thinking, any process associated with the Shuttle deserved the same high level of attention and care.

One day during a discussion, a simple observation was presented to the group that helped change how our group looked at human error failures. It was pointed out that a major concern for a rancher is to prevent his cattle from wandering off his land and onto highways or adjacent property. To keep this from occurring a rancher does not study his cattle trying to figure out why his cattle would want to leave a safe field full of green grass and go out on a highway. The rancher does not try to determine if his cattle are happy, intelligent, or properly motivated. A rancher does not try to analyze what is going on inside a cow's head. A rancher just focuses on building and maintaining his fences.

A rancher understands that cattle, by their very nature, will just wander. If he has a weak section of fence, sooner or later one of his cattle will stumble upon that weak section and go through it. To prevent this problem, a rancher just properly builds and maintains his fences. He will always have some maintenance routine where he inspects his fences and repairs or rebuilds them when weaknesses are observed.

Our group wondered if we could deal with workforce human error better if we looked at it the same way a rancher looks at cattle wandering off through a weakness in a fence. Our group began discussing if maybe we could accomplish more by looking at human error in a different way. Maybe we were focusing far too much of our efforts trying to better understand the "cattle" when what we really needed was a better understanding of our "fences".

We realized that when we examined any Shuttle processing task there were always some controls in place that functioned to make sure the task would be worked exactly as how the engineers responsible for the task wanted it to go. If you thought of these controls as being a type of "fence" built by the process owning engineers to contain the actions performed by the workforce, then it was clear to see the similarity between the engineers and their task controls and a rancher and his fences.

When an engineer wants a task performed a certain

way he might choose to write a procedure which he expects the technicians to follow exactly. That procedure is a control just like a fence. When we put a technician through a training class to instruct him on how a task needs to be performed in a certain manner, that training class is also a control. Behind every action being taken by our workforce there was some type of a control whose purpose is to assure the process is worked as the engineer or anyone else owning the process intended. If controls are a key part of doing a task correctly, then clearly they also have a significant role in tasks that are worked incorrectly. We had always viewed human error as a situation where an employee fails and therefore lets the company down. Now we were wondering if maybe we should be thinking of human error as more of a situation where our processes fail and let the employee down.

The group felt that a better understanding of the controls we were using in Shuttle processing might prove valuable in understanding why our workers made errors. Since our group had all this data on thousands of human error process failures, maybe we could analyze that data to see if there were any relationships between certain kinds of controls and human error process failures.

6 - UNDERSTANDING OUR CONTROLS

We figured the very first thing we needed to do was to create a listing of all of the major controls that applied to our workforce on a regular daily basis. Being that we were processing something as complicated as the Space Shuttle, with so many of the tasks being absolutely critical for a safe and successful flight, we assumed there could be a hundred or more different types of regularly used controls. What we observed completely surprised everyone.

Our manager called his entire group of Corrective Action Engineers into our conference room and informed us that we were going to spend the entire day, if needed, trying to list all of the common controls we use on a regular basis. Our manager created a spreadsheet to capture the list of controls and then he went around the room asking all of us to start calling out controls while he entered them into the spreadsheet.

This listing of the common controls did not take the entire day to develop. We were shocked when we realized that we had completed our list in about 20 minutes. We all looked at each other and thought we surely must be missing something. How was it possible that we were processing arguably the most complicated machine ever built using only a little more than a dozen types of controls on a daily basis? We looked over our list several more times, but each time we came up with the same very limited number of common controls.

- Written procedures with the person performing the work buying off that the task was properly performed with or without testing
- Written procedures with various combinations of buyoffs required (Tech, QC, NASA QC, Systems Engineer, etc.) stating the task was properly performed with or without testing
- Warning Placards / Maintenance Placards / Operating Instructions Placards
- Training with Competency Certified
- Non-Certified Training
- Desk Instructions
- Workforce Standard Procedures
- Bulletins
- Tailgate Meetings
- Software Controls
- Mechanical Controls / Lockouts
- Direct On-Site Real-time Instructions

Once we had created our list of common controls, the group then began evaluating our historical database on workforce human error. We pulled information on which controls were being used when the human error originally occurred and what additional controls had been added as part of our corrective actions or at NASA direction. The database on the thousands of investigations we had performed over the nearly 20 years the group had been in existence provided us with the tool necessary to determine which controls consistently worked well and which ones could not be counted upon to entirely prevent human error and its consequences. We understood intuitively that all controls were not equal. It was obvious from the start that the control of a bulletin telling the workforce to "Be careful before turning on a circuit breaker to make sure no one is working on that same circuit" is not nearly as strong a control as a circuit breaker that is mechanically locked out with the individual who could be potentially electrocuted having the only key to the lock. We decided to use our database to see if we could break down the controls we were using into some type of ranking.

We cleaned up our original list and broke some of the controls down into more detailed descriptions. Comparing our investigation database findings against our list of controls, we were able to list the controls based on how many times the control failed while taking into consideration how often the control was used. We came up with what could be considered a ranking of our controls based on their overall relative strength. When the following controls were used during Shuttle processing the control ranking was determined to be as follows:

1. Mechanical Controls / Lockouts
2. Software Controls
3. Written procedures with various combinations of buyoffs required (Technician, Quality Control inspector, NASA Quality Control inspector, Systems Engineer, etc.) verifying the task was properly performed with final product inspection and testing.
4. Written procedures without buyoffs combined with independent final product inspection and testing.
5. Written procedures with buyoffs, without independent final product inspection and

testing.
6. Written procedures without buyoffs, without independent final product inspection and testing.
7. Warning Placards / Maintenance Placards / Operating Instructions Placards
8. Direct On-Site Real-time Instructions
9. Training with competency certified
10. Bulletins
11. Tailgate Meetings
12. Training without certification
13. Workforce Standard Procedures or Desk Instructions
14. Posters, Safety / Quality slogans

This order of relative strength was based on our database evaluation, but we wondered what were the specific traits of a control that moved them up or down the ranking. What we found was that the very best controls, Mechanical Lockouts / Software, when properly implemented completely removed the potential for the worker to even make the mistake. You can't send a bad command if the software won't allow it and you can't flip a breaker if it is locked down. Not a very surprising finding.

The next most successful groups of controls accept the fact that a human can still make an error in this process even with clear work step instructions located directly where the actions are taking place. An error can still happen to any individual, but these controls require a second, or even third, individual having equal or better skills and knowledge to also independently assure the task was correctly performed. With these controls, the worker also takes direct responsibility for performing the task by identifying the step with his personal signature or stamp. These controls might also involve some level of testing that the work was done properly. These controls work well because it is extremely rare for several highly skilled individuals to all independently make the exact same error.

With the next, slightly less successful, grouping of controls, the directions for the desired actions are also clearly located where the actions are taking place. The person performing the actions also takes direct responsibility and assures each step has been accomplished by signature or stamp on each step. These controls rely only on a single individual not to make an error. If that individual fails there is no inspection or second opportunity to catch the mistake. It is not uncommon to find human error process failures occurring when these type of controls are being used in a process. These controls fail because as a simple fact, workers are capable of making a mistake even when they know exactly what to do.

The next group of controls, when properly implemented, still resulted in a human error on a regular basis. These controls expect a worker to remember some instructions that were previously told to them, maybe several months previous. Workers are humans, and humans forget things. When a process assumes a worker will not forget something is when errors are going to occur.

The last group of controls is really just a shot in the dark hoping that they may provide some benefit. These controls are based on the idea "it doesn't hurt and it might just help". These controls are just as likely to fail as they are to succeed, but they are still better than nothing. Since they are relatively inexpensive to implement, they are often chosen when a true fix to a problem cannot be determined. Rather than just doing nothing, this last group of controls is what typically gets implemented.

What is very important to realize is that these controls and their order of ranking were solely based on the way our own particular workforce was already functioning. Other companies performing other types of tasks may use a different set of controls and their order of ranking could be totally different. For example, a home construction business may have direct real-time on-site instructions as their number one strongest control with tailgate meetings as their second strongest control. Every company is different, so their control selection will also be what fits best for their applications.

The simple way to look at controls is to always remember that humans can forget things, they can get confused, and they can get distracted. When working with a human workforce these weaknesses have to be taken into consideration. The best controls address each of these weaknesses. Workers can forget things, so the best controls have the required actions right there in front of them with nothing to remember. Workers get distracted, so things like checklists and step buy-offs keep important actions from being missed. Workers get confused, and to address worker confusion, the best controls typically involve clear and detailed work instructions. Independent verification of correctness or final product inspection and testing addresses all three potential problem issues of confusion, distraction, or failing to remember. Any process that does not address the potential for workers to forget things, get confused, or get distracted can be expected to eventually fail due to a human error problem.

The move away from the human part of the problem and the focus on the process itself made sense to us. We understood that it was beyond our abilities to fix humans, but not all that difficult for us to fix processes. Controls were the key tools that we were counting on to assure that tasks were properly performed and yet there had never been any analysis done or even data collected specifically on controls that would have provided insight on which controls we should be relying on and which controls were a waste of time for us to even put in place. Not only was there no analysis ever performed, we could not find where anyone else had previously made a list of what common controls we used on a daily basis.

We were working on contract to NASA and one thing that NASA insists on is metrics on every important activity we perform for them. NASA always wanted to be able to spot the beginnings of negative trends long before serious problems began occurring. The fact that NASA Quality managers had never asked us for any data from our investigations that focused specifically on controls and how they were performing was an indication to our group that we were working on something that might be different and potentially very useful.

7 - CONTROLS AND PROCESSES

Let's consider a hypothetical situation where a hydraulic valve needs to be rebuilt. The individual responsible for accomplishing this objective is the owner of this process. This individual is very knowledgeable of exactly what the hydraulic valve build process needs to be and he has his own clear plan/design of what must be done. The next question is what kind of controls does he use to relay the actions required per his plan to those personnel who will be working the process in order to assure that it gets worked exactly as he desires it to be done.

Consider a situation where the process owner places the o-rings, pistons, springs, fittings, and the tooling needed to build the hydraulic valve on a table in a room. The process owner then goes out on the street and approaches the first person he sees and brings him to the room where the parts are located. The process owner then sits down with the person from the street and spends a few minutes explaining how the hydraulic valve should be built and then leaves the person in the room by himself to perform the work. The process owner walks away and assumes the person he found in the street builds the valve exactly as the process owner intended.

This would be a situation where the proper plan or design of what needed to be done was clear to the process owner and the process control the process owner chose to assure the valve was built per plan was the verbal instructions given to the individual off the street doing the actions. The verbal instructions were totally accurate and this is a process that is potentially capable of getting the hydraulic valve built exactly as the process owner desired.

Now let's take this same situation, but instead of just giving verbal instructions and leaving the guy off the street alone inside the room to perform the action, the process owner takes additional actions to assure the valve is built as desired. In this second case, the process owner puts the individual performing the work through a very intense training class on building hydraulic valves. The class teaches the hydraulic valve build process exactly as how the process owner's design would want it to be performed. The worker spends two weeks in this class and completes the training by passing a certification exam on hydraulic valve building.

At the end of the training class the process owner places him in the very same room and tells him to build the valve, but in addition to the training, the process owner also has provided detailed written work steps to be followed and to be bought off by the worker to assure that nothing taught in training is forgotten or missed. The process owner also arranges for another person who is very knowledgeable of hydraulic valve building to watch each step and independently verify that the written steps are followed. Finally, the process owner has a piece of equipment built to automatically test the finished hydraulic valve to assure that no mistakes were made during the building process.

These are two examples where in each case the

process owner had the very same goal and knew exactly what fabrication steps the person building the hydraulic valve needed to perform. In each case, the directions provided were accurate and the process was potentially capable of resulting in a properly built valve. The difference was that in the first example, the process owner used a weak control to implement his desired process that was vulnerable if the person working the steps forgets a step, becomes confused, or gets distracted. In the second situation, the process owner designed a process using very strong controls to assure the desired actions will ultimately be performed correctly, even if the person performing the work forgets a step, becomes confused, or gets distracted.

These two examples show that while both situations were capable of producing a properly built valve, only the second example could deal with typical potential human error situations. It is easy to understand that processes using very weak controls commonly fail while processes using very strong controls typically work well, but what about all of the controls and combinations of controls in the middle ranges? Processes being directed by strong controls are usually more difficult and expensive to put in place, while processes being implemented by weak controls are usually less costly and easier to implement, but far more likely to result in a human error failure. How do you choose the right control, or a combination of controls, for any particular task? If the hydraulic valve was only going to be used to raise or lower the cutting deck on some yard lawn mower, maybe the weak control used in the first example was actually very appropriate. If the hydraulic valve was going to be used to unlock the landing gear doors on a commercial passenger aircraft, maybe the strong controls in the second example were still not strong enough. Our group decided that we needed to focus our actions on developing a better understanding of the relationship between process controls and process failures due to human error.

Our group now knew what controls we had to work with, what their relative strength was, and the possible rationale for why strong controls worked when weak controls failed. At this point, it would seem the simple answer to eliminating human error failures from all of our processes was to use only the strongest controls and make sure they were properly implemented. A nice thought, but clearly not practical or cost effective.

The group decided the very best solution was to identify tasks that had the potential for a serious consequence if a human error occurred and assure that in those situations they were using strong controls that were being properly implemented. For Shuttle processing, these would be human errors that could result in consequences like death, loss of vehicle, serious injury, loss or serious damage to critical hardware, loss of mission, or significant delay in processing. For these high consequence processes, we needed to assure we were using the very strongest controls and to also make sure they were being implemented properly. For other processes that had lesser consequences from a potential human error problem, we could go with controls that were cheaper and easier to implement.

The downside of using controls in this manner is that human error process failures were going to continue to occur occasionally in low consequence processes. We were going to have to convince management and our NASA counterparts that finding human error process failures in low consequence processes that were using weak controls was not indicative of any overall workforce problem.

While all human errors are undesirable, the reality is that strong controls are typically very expensive to implement. Our group continued to document every incident where a human error occurred, but we worried less about where a weak control failed on a low consequence task. As long as there is a low consequence potential, then it just made sense to utilize a lower cost weak control and live with the potential of human error when it is cost effective. There is no logic in having a thousand dollar control in place to prevent a ten dollar consequence.

The things our group began to mainly focus on were;

1. **Investigate all failures of strong controls regardless of whether the failure had actually resulted in a negative consequence or not**.

Our group became absolutely driven to completely understand every strong control failure. It seemed that historically, whenever a human error occurred, the common corrective action was to put an inspector buy off on the step. NASA also insisted on having one of their inspectors buying off on nearly every action regarding flight hardware regardless of the potential consequence of a human error during the process. This resulted in a significant amount of strong controls being used where the consequence was low. Our group started working to a new set of priorities that we believed used our limited resources much more efficiently. To our group, the actual consequence that resulted from the human error was not nearly as important as the strength of the control that failed. Strong controls always had to work and they needed to be counted upon in a process without exception. Previously, the criticality of the hardware or the extent of damage from a human error was what determined the level of investigation performed. Where a serious consequence did not ultimately occur as a result of the human error, there previously had been only a minimum investigation performed. Now to our group there was nothing more serious than the failure of a strong control regardless of the final consequence that resulted.

2. **Identify any process where a human error had the potential for a serious consequence but did not have a strong control preventing the error**.

We found many instances where existing processes were trying to prevent a serious consequence using only one or more weak controls. These situations were looked upon as hidden dangers that were not being considered previously. It was the simple processes that were common and considered relatively harmless that often resulted in the most serious consequences. Falling off a ladder will kill someone just the same as a Shuttle main engine exploding. Engineering efforts were always more focused on preventing problems with high visibility flight hardware, but simple things can have serious consequences as well.

3. Identify processes that could be upgraded to a stronger control without increasing costs.

We found during investigations that many of our controls had the potential for some improvement. All human error is undesirable, even those with a low consequence. Some stronger controls were actually less expensive to implement than some weaker controls being used. We decided that if a control could be easily upgraded to a stronger control in a cost-effective manner, then we would try to do that.

4. Look for cost savings for NASA by identifying where expensive strong controls were being used unnecessarily to address minimal consequence situations.

The strongest controls are typically the most expensive controls to use. Our group saw the potential for major cost savings by always matching up the lowest cost control that was suitable for the potential consequence. Our NASA counterparts that the group dealt with on a daily basis especially liked the idea that this new way of looking at human error through control strength could also be used as a logical way of reducing processing costs without increasing risk.

8 - PROCESS DESIGN BASICS

For many months, our group performed workforce human error investigations while considering the new direction that when human error occurs, it is far more important to focus on the process itself and the controls being used rather than the human involved. We gathered a lot of data on what we were seeing and while analyzing this data we observed something very troubling to us. The data showed that processes using strong controls significantly reduced the number of human errors, but they did not eliminate them entirely. Some tasks that employed very strong controls still encountered problems from human error. Clearly the fact that strong controls were employed was by itself not enough to eliminate the potential for human error problems.

In order to better understand what was happening, the group went back to analyze what our database was now telling us. We studied every human error process problem and found that there were certain types or families of flaws being observed. We placed each failure into groups based on how the process actually failed. When the process failed as a result of a human error combined with a weak control, those were entered into the "Failed Weak Control" category. When there was a human error process failure that was using a strong control we placed them into the "Failed Strong Control" category. We eliminated all of the Failed Weak Control errors from further evaluation because when weak controls are being used, you have to expect that human error failures will periodically occur within the process. We were further able to place all issues in the Failed Strong Control category into one of five smaller groups. These smaller groups were as follows:

Define – The process as designed was flawed when it came to being perfectly clear on exactly what actions were required to be taken at a specific instance.

Assign- The process as designed was flawed when it came to being perfectly clear on exactly who needed to be performing a specific action. Responsibility for taking specific actions was not clearly assigned.

Train – The process as designed failed to assure that the individual assigned with performing the actions had all the required skills necessary to perform the actions.

Organize – The process as designed failed when any of the supporting processes and activities necessary to perform the required actions were somehow lacking and contributed to the human workforce error. Examples of these supporting activities include: tool calibrations, lighting, work area arrangement, tool selection, scheduling, communication, etc.

Monitor – What our group observed was that processes don't typically just go from being great one day to failing the next day. There is almost always a period in between where the process starts to break down a little and begins showing indications of problems. Different things can lead to the beginning of the breakdown such as key people retiring or getting transferred, tooling gets changed, new requirements get added, or new skills become required. A process that is presently working fine and has properly implemented the elements of Define, Assign, Train and Organize can break down when any changes occur. An ongoing means of management monitoring the processes that are capable of catching problems early is necessary to maintain confidence in any process. Problems that should have been caught with closer management.

We began to refer to these five elements of the process design (Define, Assign, Train, Organize, and Monitor) simply as DATOM. What the group realized, was that strong controls can only assure that the process gets implemented exactly as designed. If the process had a design flaw (it fails one or more of the DATOM elements), then in those situations strong controls will only assure the process gets worked with the flaws included. What now was apparent was that strong controls could not help a process with flaws. Strong controls combined with a flawed process can actually make things worse.

9 - MAKING SOME PROGRESS

Our group went back to our database and looked at those processes affected by human error failures. What we wanted to know was how many times a process that was properly designed (no DATOM flaws) and used strong controls experienced a human error failure. We found that out of the thousands of human error process failures we evaluated, there was only a small group that could not be explained by either weak controls or a flawed process design (DATOM).

This was an important finding for us. This meant that we were very close to realizing a simple way for identifying when a process was likely to encounter a human error failure and when it was likely to succeed in avoiding human error failures. The only thing remaining was to address this small remaining group that did not fit the DATOM failure or weak control rationale for failure.

We decided to break a basic process down into its key components. The group arrived at a consensus that all processes seemed to have these same elements –

The *Process Owner*: this was the person who was responsible for understanding the final goal and tasked with designing the process that would accomplish that goal. In reality, the *Process Owner* can be thought of as any individual in the process with approved permission to change the process. The ultimate Process Owner of any process is the last person who was allowed to change the process.

The *Process Worker*: this was the person or persons who were responsible for following exactly the actions prescribed by the *Process Owner* when working the process. If a *Process Worker* has approved permission to change the process he is working, then he also becomes the *Process Owner*.

The Process Design: the *Process Owner*'s plan for how the process needs to be worked in order to reach the desired goal.

The Potential Consequence of Process Failure: the potential realistic worst-case result of a human error occurring within a process.

The Process Controls: the *Process Owner* chooses the suitable controls that are to be used to assure that his chosen process design gets relayed to the *Process Workers* for implementation. Controls come in different strengths. The strongest Process Controls are those most likely to assure that the Process Design gets implemented exactly as the *Process Owner* desires. The weaker the Process Control, the more likely the Process Design will not be implemented as desired by the *Process Owner*. A strong control might be a written procedure with quality control (QC) verifications, while a weak control might be a wall poster.

When our group looked into the data on strong

controls and human error process failures, what we saw was that far and away the majority of failures were the result of using weak controls or poorly designed processes (DATOM flaws). The remaining failures that did not fall into those two groups were found to be caused by confusion over who actually owned the process. This ownership confusion led to unapproved changes being made to the process resulting in human error situations.

An example in our database of a failure that was the result of confusion in process ownership was where a technician was working to clearly defined steps in a written work document. The process was well designed (no DATOM flaws) and the control strength was strong and suitable for the potential consequence of a human error. The problem began when a technician came to a step in the work document that might have caused a work stoppage if the step was followed exactly as written. Rather than following the work document instructions and possibly having a stoppage, the technician worked the step differently so that the potential for the work stoppage was avoided. In this situation, the technician knew exactly what the engineer (Process Owner) had wanted to be done, but felt that he would be allowed to do something different to avoid a work stoppage.

When the *Process Worker*, without *Process Owner* approval, works a process differently than the designed plan prepared by the *Process Owner*, that *Process Worker* is in effect taking over ownership of the process. Our group considered this to be a very dangerous situation even though there was no negative impact in this particular example. In fact, some in management felt this kind of "initiative" on the part of the technician was very desirable since an expensive work stoppage had been prevented by his actions. My group had strong disagreements with some in management whether an incident such as this should result in an award or disciplinary actions for the technician involved.

This instance highlighted just how easy it is for an organization to allow confusion on process ownership to take hold. Confusion on process ownership can lead to very dangerous situations if not adequately addressed. Our group had to argue with management on many occasions that these types of actions should not be allowed. Management had to make clear to all personnel that when the process cannot be worked as written, the process must be stopped until the *Process Owner* provides different direction. Even if the alternative actions might be beneficial, only the established *Process Owner* is allowed to modify the process.

Process Ownership can best be described as how well does an organization do in enforcing compliance with process directions. A *Process Owner* can do a terrific job in designing his process and he may choose the strongest of controls to implement his well-designed process, but it is all for nothing if an organization does not assure that the *Process Workers* comply with the process. If an organization allows exceptions to complying with a *Process Owner's* designed process, then you really do not even have a process at all.

Mishap investigation will too often start out with the *Process Workers* explaining how they were not following the procedure exactly as written because something unplanned occurred. They were behind schedule and rushing to avoid darkness. The weather was turning bad, so they changed the process to get done more quickly. Some key individuals were out sick or on vacation, so they had to change the process to work around them not being there. Some equipment broke down, so they were working around not having that particular piece of equipment available. When processes are not worked as the *Process Owner's* plan directed, it can quickly become a very bad situation. Management needs to be made aware whenever processes have not been worked as the *Process Owner* intended and make sure these situations are understood and actions taken to limited them in the future.

10 - ALL HAIL THE PROCESS OWNERS

It cannot be stressed too strongly that the role of *Process Owner* must be established and clearly understood that no one else has permission to change the process from what is being relayed through the process controls. There can be no confidence in any process where process ownership is not strictly enforced. It is the *Process Owner* who is best informed as what the final goal is, and it is the *Process Owner* who is ultimately responsible if that final goal is not achieved.

The military understands that when an officer gives an order for an immediate attack on a specific location that, without question, an immediate attack is exactly what has to occur. In the military, you would not even consider the possibility of soldiers taking the "initiative" and on their own deciding to delay the attack until tomorrow when the weather might be better. *Process Workers* cannot be allowed to modify the actions that have been specifically directed by the *Process Owner* through the process controls. Likewise, there should always be an approved means already in place that has been created by the *Process Owner* where *Process Workers* can call a "Time Out" if there is confusion regarding exactly what actions the *Process Owner* wants performed or if a safety issue arises. There is a time for "showing initiative and thinking out of the box" but that time is definitely not when the process is actually being worked. Any ideas the *Process Worker* might have for improving the process can be relayed later to the *Process Owner* for consideration for the next time the process is performed.

Our group now had a clear idea of what we needed to do to address the human error process failures that were occurring, and we had a database of thousands of human error process failures that backed up our rationale. We knew that if we could understand why a process breaks down and fails as a result of human error, we could also use that same knowledge to look for weaknesses in processes where there has yet to be a failure. Finding a way to identify a process that was likely to fail is what we had been looking for all along. Fixing something before it breaks was what our NASA counterparts wanted from us and we were getting close to having that.

We began to view processes in the most basic form as simply someone being directed to walk down a path heading to some destination. The *Process Worker* is the person walking the path and he is heading to some final destination that the *Process Owner* desires him to reach. The *Process Owner* is providing the directions that are intended to assure the *Process Worker* will arrive at the desired final destination. Along the path, there are numerous forks and at each of these forks, the *Process Worker* has an option as to which direction to take. The role of the *Process Owner* is to recognize where each of these forks will occur and provide the *Process Worker* in advance with the desired direction the *Process Owner* wants the *Process Worker* to take. The manner in which the *Process Owner* chooses to provide the desired direction is the Process Control. If the *Process Worker* should ever take a turn along the path that is different than what the *Process Owner* had desired, then this would be considered to be a human error. If the *Process Worker* continues down the wrong direction and something harmful occurs as a result, you then have a negative consequence from a human error.

Human error is just a situation where the *Process Worker* goes in a different direction than what the *Process Owner* had intended him to go. There is no bad guy in this situation, no villain, just someone taking a different path than what someone else expected them to take. This situation often happens when a fork or decision point along the way is encountered that the *Process Owner* failed to anticipate and therefore no direction was provided. Without clear direction, the *Process Worker* made the choice on his own and he chose incorrectly. Even when a *Process Owner* correctly anticipates a fork or decision point and provides good direction on which direction to proceed, the *Process Worker* may still head down the wrong path. This happens because *Process Workers* are human, and humans get distracted, they forget things, and they get confused even when good directions have been provided. This is why it is so important for the *Process Owner* to recognize potential high consequence situations and choose the strongest controls available for those situations.

The Process Control is whatever means is chosen by the *Process Owner* to provide the *Process Worker* with the desired direction to take whenever the path encounters a fork and there is an option to go a different direction. The consequence of taking the wrong direction when a fork is encountered, will determine the type of control to be used. If the consequence of going the wrong direction at a fork is minor, then the Process Control may just provide nothing more than simply stating the desired direction to take. If the consequence of making the wrong choice is more serious, the control might provide the desired direction, plus something additional to identify the mistake in a timely manner to avoid a problem or limit the damage (QC verify, final product inspection, final product testing). If the consequence of a *Process Worker* taking the wrong direction is critical, then the Process *Owner* might choose a control that assures the wrong direction cannot be taken (software override, switch locks, QC witness).

Being a *Process Owner* is very difficult. They not only have to provide good directions, they also have to identify all potential decision points where a problem might occur. A *Process Owner* also has to determine what controls to use based on the consequence of a human error. During the Space Shuttle program, the systems engineers were the ones that gave directions, typically through written procedures. The systems engineers (*Process Owners*) that could put out the best-written procedures would encounter the fewest problems when the technicians (*Process Workers*) were performing the procedures. A process that is well designed initially prevents serious problems later on when the process is worked. All organizations should understand the importance of good *Process Owners*. There is a great deal of pressure on *Process Owners* because they are the ones ultimately responsible for the success or failure of a process. Good *Process Owners* are extremely valuable to any company and they should be regarded highly and well rewarded.

Our group realized quickly on that a *Process Owner* could generate a process more quickly and easily by incorrectly assuming that the potential for human error does not exist for their specific process. A *Process Owner* can always justify putting out a process that does not address the potential for human error by convincing themselves that "this is why we hire good people" or, "as long as everyone is careful, there should not be a problem", or even "we just need to have more discipline and make sure the technicians know mistakes will not be tolerated". My group would often point out places in their process where the possibility of human error should be considered and the potential addressed. It was common to have a difference of opinion, and often our suggested changes would not be implemented. It is very difficult to argue the position that a human error could occur when there is no previous history of human error in that process. In many instances, the complicated way changes got approved on the Shuttle Program would make it extremely difficult for the *Process Owner* to make any modifications.

The Shuttle was made up of components designed and built by many different companies under contract to NASA. Even though my company on contract to NASA supplied the workforce that prepared the

Shuttles for launch, we did not own responsibility for the actual hardware. Our engineers were the *Process Owners* when it came to providing work instructions to our technicians, but they were limited by having to obtain concurrence for any process change from all companies involved when the process dealt with hardware designed and built by other companies. It was so difficult to obtain all the required concurrences, that often the *Process Owners* would not even attempt to take on the challenge to change their processes, even if they agreed with us regarding the process flaws and weaknesses we were pointing out.

I have a workshop at home in my garage. I know that I am capable of making a mistake that could cause a fire while working in my shop, so I have a fire extinguisher nearby. The fire extinguisher costs about $35 and I replace it every 5 years. Now if a fire extinguisher cost $2000 rather that $35, I would probably not have a fire extinguisher there and I definitely would not be changing it out every five years. I would probably just say to myself that I am a cautious guy who can get along without the fire extinguisher by just being more careful to avoid a fire when working in the shop. I would just be kidding myself, because I know I would act exactly the same way whether there is a fire extinguisher nearby or not. So in this situation, I am just like any other *Process Owner*. When it became too painful (expensive) for me to properly address the potential for human error and have a fire extinguisher nearby, I merely went with the "feel good" excuse of just stressing the need to be extra careful. As the *Process Owner* in this expensive fire extinguisher situation, I avoided taking actions that were too painful to implement, even though I knew the potential for me to cause a fire in my shop still existed.

Our group had to work around the fact that for some *Process Owners*, it was just so difficult to change their processes that they were not going to consider making a process modification based on just the mere potential for a human error. In these difficult to change processes, the *Process Owners* would only take on the effort of changing their processes when a human error had already occurred or there was an existing history of a problem. The reality was that our group was able to identify processes that had flaws and weaknesses regarding the potential for human error, but often those flaws and weaknesses would only be addressed by *Process Owners* who could easily change their processes. An organization typically believes that requiring many individuals to concur with any changes makes the process more resistant to errors. What we observed was that when you make a process too difficult to change, it can also limit the ability to take preventative actions. When it becomes too painful to take the steps necessary to prevent problems that might only possibly happen, people will just become overly optimistic and assume the problem will never happen in their situation.

When a human error mishap has already occurred, it is very easy to get everyone to agree to make process changes that will prevent a recurrence of the problem. Getting those same people to agree to any process changes before a mishap has actually occurred is a far more difficult task. Our group realized that if we were ever to comply with our NASA directive to increase preventative actions, we were first going to have to develop a means of convincing others of potential problems that were not based on an already existing history of problems.

11 - CAN'T FIX BEING HUMAN

Within a short period of time, our group began to get very adept at using Process Ownership, Process Design (DATOM), and Process Control Strength evaluations as a tool in human error problem investigations to identify quickly and accurately why the process broke down and failed. Once our group accepted these basic concepts, we realized that it became a rather simple matter to identify why a process encountered a human error failure. Our manager decided that we would call this three-part investigation method a Control Based Risk Analysis (CoBRA). We all agreed to just simply refer to this analysis as a CoBRA.

Before long, our group transitioned from wondering why a worker made a mistake to just accepting the fact that everyone is capable of making numerous mistakes on a daily basis. For people who work in offices, we dial wrong numbers, we forget names, we send e-mails to wrong people, and we get confused on which meetings we were supposed to attend. Our list of potential mistakes goes on and on. Yet we somehow expect technicians working with complicated equipment to never commit an error.

If you were tasked with designing a building and told you had to work with beams that were supposed to support 10 tons of weight, but on rare occasions one might fail at only 8 tons, would it not make sense to just design the building assuming that all beams have the potential to fail at 8 tons? Knowing that some beams have the potential to fail at 8 tons, would anyone actually decide to design a building where all beams must support 10 tons and just do a mishap investigation each time the building collapses because a beam fails at 8 tons. Why then is it when it comes to processes involving humans, that is exactly what we do? We know humans forget things, get confused, and get distracted but we still build processes based on the assumption that humans will work the process without mistakes. If we accept that humans have the potential to forget things, get confused, or get distracted why are we not building our processes based on the assumption that this can happen?

Our group often used an airline pilot as an example where a human error failure has extremely serious consequences. Airline pilots are some of the most intelligent, highly trained, well compensated, and closely monitored individuals of any workforce. When it comes to motivation to do things correctly, what better motivation is there than the knowledge that they too will die in a crash resulting from a human error on their part? So then why is it all too common to hear of planes crashing as a result of pilot error? The simple explanation is that pilots are human and all humans forget things, get confused, and get distracted. If we cannot completely eliminate the potential for someone like an airline pilot from making a human error, then how can we not accept as fact that everyone has the potential for making a human error at any time? The potential for human error always exists and all processes should be built in a way that accepts that simple reality.

A process designed with this human error potential in mind does not assume there will not be mistakes, but rather accepts that the potential is always there and deals with the fact that mistakes are going to eventually happen. Looking at the airline pilot again for an example, the pilot should know how to configure his plane for takeoff since he has done it thousands of times, but there is also a checklist to follow in case he would make an error and forget something. Understanding that an individual may still miss an item on a checklist, both pilot and co-pilot work the checklist together with each typically having their own copy. The pilot must also keep his plane at a safe altitude, but there is also an alarm that signals close proximity to ground in case he makes an error with altitude. The airline pilots should always be looking out for other planes, but on larger aircraft, there still is a collision avoidance system to tell him to climb or descend if a mistake is made and his plane gets too close to another plane. An airline pilot is trained to fly in a manner that prevents a stall, but there is still an alarm that goes off if he makes an error and his plane approaches a stall. Pilots also make many errors that are caught by alarms, software, or co-pilot intervention that no one is ever aware of. Planes continue to have crashes that were the result of human error on the part of the pilot.

The checklists, ground proximity alarms, collision avoidance systems, stall alerts, and other safety controls all came about because pilots have made mistakes in the past and planes crashed with people dying. They came about after pilots made mistakes that were originally assumed they would not make. This sadly repeating cycle of mistakes first with fixes later is the result of the false belief that intelligent, well trained, well-motivated people do not make mistakes.

To my group's previous way of thinking, when mistakes did occur the person making the error was viewed as somehow different and inferior to everyone else who had worked the same process previously without making that mistake. This cycle of failures, blaming and then fixing seemed to always work the same way for us. Someone makes an error in a process that results in some problem. There is an investigation done and we find that the person who made the error should have known better due to some previous training, procedure, tailgate meeting, bulletin, or some other form of instruction that the individual had received. Previously to us, it was as if a crime had been committed and we were detectives whose job it was to gather evidence for a trial. The person who made the mistake was typically blamed and usually given at least some type of reprimand. A suitable "fix" was then implemented such as an additional QC buyoff step, tailgate meeting, or enhanced training. It was always the same cycle and the result was we were forever chasing the problems and never getting ahead of the problems.

Our group decided that we would change how we did our investigations. The group decided that we were no longer going to focus on the human error itself, but rather on the process during which the error occurred. We were going to become more knowledgeable on process design. We were going to learn how to be better at building and maintaining controls. For future investigations, a workforce error would be looked upon as nothing more than a situation where the process, for some reason, did not go as the *Process Owner* intended. If a serious negative consequence resulted, it was because the process failed to adequately address the potential for that serious negative consequence. When you consider the undesirable situation as a process problem, rather than a human mistake, it actually makes the investigation much simpler. To us, the problem became less about a human making a mistake, and more about a process that was unable to identify and deal with potential human error problems.

We began to view the person making the mistake as more of a victim of an inadequately designed or implemented process rather than the actual source of the problem. If the consequence of a human error in a process is potentially very serious, you have to take into consideration the possibility of human error when you are designing the process. There is a good reason why the Boy Scouts motto is "Be Prepared" and not "Pray for Good Luck".

12 - KEEP IT SIMPLE AND QUICK

It did not take long for our group to transition from only being able to chase worker mistakes after they already happened, to recognizing processes where human error problems were likely to occur. We went from wondering why a worker made a mistake to accepting the fact that everyone is capable of making a mistake and it is the process itself that is the major factor in determining if a human error even happens and a negative consequence results.

We no longer thought of human error as being a people problem, we now considered human error to be a simple situation where the *Process Owner* expected the process to be worked a certain way and that did not happen. Human error was merely a situation where the process broke down in some manner. We now began our investigations with the goal of determining why the process failed, not why did someone screw up.

We came up with several guiding principles we accepted as a basis for our investigations:

1. There can be no confusion on who owns a process. When multiple individuals are permitted to change a process, then there is no clear ownership of the process. A lack of clarity of ownership can result in confusion regarding actions to be taken. The entire process design (DATOM) falls apart if it can be changed during the process by anyone other than the identified *Process Owner*. Any confusion regarding

Process Ownership creates a definite potential for a human error failure in the process.

2. When a process fails as a result of human error, the focus needs to be mainly on Process Ownership, Process Design, and Process Control Strength and not the individual involved. Local management knows best what actions to take involving the individual. Local management will protect the good workers or use the incident to remove undesirable workers. We had to remember that often many other factors were being considered by local management besides just this particular incident.

3. An easy and quick way to see if a process is well designed is to do a DATOM check. If Define, Assign, Train, Organize, and Monitor are analyzed and they are OK, then you can accept that the process is designed properly.

4. Controls are the means by which the *Process Owner's* plan of how the process should be performed is conveyed to the *Process Workers*.

5. Strong controls are very good at assuring that the personnel working the process perform it exactly as the *Process Owner* designed it.

6. Weak controls do a poor job of assuring that the individuals that are working the process perform it exactly as the *Process Owner* designed.

7. Human error is highly unlikely when:
 a) There is no confusion over process ownership.

b) The *Process Owner* does a good job of designing his process (passes DATOM).
c) Strong controls are used to assure personnel working the process are doing exactly as the *Process Owner* desired.

8. Human error is likely to occur when:
a) There is confusion over process ownership.
b) *Process Owner* does a poor job of designing his process (fails DATOM).
c) Weak controls are used to assure the process is worked exactly as desired.

The group realized that by just focusing on these specific concerns, it became rather easy to identify why a process encountered a human error failure. We began to develop a kind of quick checklist to go through when looking at a human error process failure to determine what went wrong. We came up with the following:

1. Was it clearly understood who was the *Process Owner* and who were the *Process Workers*? Did everyone concur that the directions provided by the *Process Owner* were the only acceptable actions that could be taken, or were they viewed as only strong "suggestions"? What exceptions are allowed for not following the *Process Owners* directions?

2. Was the process design (DATOM) satisfactory?

3. Was the control strength being used to implement this process suitable for the realistic worst case

consequence of a human error failure in this process and was the control flawed in any way?

The key to any investigation is to know what to look for and what questions to ask. Everyone in the group began going down the same path in our investigations and it was remarkable how quickly everyone caught on. By following this mental checklist, it was simple to ask the right questions. By concentrating on getting information that pertained to these specific issues, the group was able to quickly arrive at what truly went wrong in the process that led to the human error failure and determine what exactly needed to be fixed. With everyone asking the same questions and looking for the same potential problems investigations were completed much faster than before. We kind of felt like we were doing investigations in an assembly line fashion.

Our group meetings where we used to status our ongoing investigations became so much more effective when everyone looked at investigations the same way. Prior to using this new method, it used to be common for our group members to say that they were at a standstill in their investigation and ask the group for advice on how to proceed. With the new process, no one ever came to a standstill because they always knew what to do next. Everyone already knew the questions the investigations needed to answer. When you knew the answers to those questions, you knew what flaws and weaknesses in the process enabled the human error failure to occur and also what was needed to fix the process to prevent a recurrence.

It was common for NASA and my management to want to address human error problems by adding an additional layer of controls. Things such as more QC inspections, or more engineering oversight with more concurrence signatures being required for any actions being taken. The corrective actions our group typically requested involved only fixing the existing process and not adding any additional controls, if the present controls already had suitable strength for the potential consequence. The only actions that we now believed ever needed to be taken, were those that would get a process to the point where it passed DATOM, had controls of suitable strength, and there was no confusion on process ownership. In our opinion, anything added to a process beyond those three elements was just adding unnecessary cost and complexity.

We used to look at every human error process failure as something unique and different because it just seemed like there was an infinite number of ways that humans could make a mistake. When we began using these techniques, what we found was that processes are easy while humans are complicated. We accepted that we were never going to understand how to eliminate the fact that humans forget things, get confused, and get distracted. When you can't make something go away, then your only alternative is to learn the best ways to live with it.

As the group became better and better at doing our investigations on human error process failures, we began to wonder how else we could use these concepts. Everyone in the group accepted that the key elements regarding human error process failures are:

Process Ownership
Process Design (DATOM)
Process Control Strength

We wondered why these same concepts could not also be used just as easily to identify an "at risk"process.

13 - GETTING AHEAD OF PROBLEMS

We believed we were heading in a direction that might lead to a way to get ahead of the human error process failures that NASA stressed to us as being so important to them. A few members of our group branched off and started to analyze some of our more critical processes. The idea was to see if they could identify problems with a critical process before there was an actual failure. The goal was to see if we could move into more clearly preventative type actions. This group knew that in a healthy process, the *Process Owner* is clearly identified, the process design (DATOM) is correctly done, and a suitable control is properly implemented. It would, therefore, seem only logical that an unhealthy process is one where the *Process Owner* is not clearly understood, or the process design (DATOM) has flaws, or there is not a suitable control that has been properly implemented. The group was trying to find unhealthy critical processes, and now we knew how to identify what an unhealthy process looked like.

We started out by taking a high consequence critical process and breaking it down into its smaller supporting processes that also needed to be performed correctly without a human error failure. For example, the large process may be installing a main engine onto the Space Shuttle. The smaller supporting processes might be how do we maintain the lifting equipment that supports the engine while we install the bolts, or how do we calibrate the torque wrench that assures the correct torque is applied to the bolts that secure the engine.

By breaking everything down into smaller supporting processes, it was much easier to determine if the process could be considered healthy or if it had potential problems. The most common process problems we observed were with process design (failed DATOM) and incorrectly matched control strength to realistic worst case consequence. Although, it was also not that rare to come across processes where there was confusion on process ownership, such as *Process Workers* making small, but still unapproved, changes.

The more we analyzed these large, very visible, high consequence processes and broke them down into their smaller, less visible, supporting processes, the more concerned we became. What we were finding was that many of these small supporting processes also could have high consequences. This was serious because these smaller supporting processes were typically using much weaker controls and being given far less attention.

It was obvious to everyone that there existed a high potential consequence if a technician installed a bolt in a wrong location when mounting an engine. What was not so obvious was the similar high consequence of the receiving inspection QC failing to assure the proper documentation of these same bolts coming from the supplier and allowing an improperly certified bolt to enter stock. It was obvious to everyone that there would be a potential high consequence if a torque wrench was set improperly when installing a critical fastener and an incorrect torque was applied.

What was not so obvious was the importance of the process that had calibrated this torque wrench and thus assured its accuracy.

Our group had previously only responded to human error process failures after the fact. We would have given a great deal more attention to a human error such as "technicians installs fastener in incorrect location during main engine installation" rather than the human error process failure "receiving inspection QC failed to verify source QC buy on section 2C on Form 1024 regarding engine bolt shipment Invoice #V7222327". In reality, the QC failing to properly verify the bolts might have been actually the far more dangerous of the two

What was also keeping many of these smaller tasks off the radar was that they often were not presently experiencing any human error failures. Often we would find processes that were very badly designed, but surprisingly, these flawed processes may have been going on for years with no human error failures having ever occurred.

We found that even the most flawed processes can still function without incident if they are being "crutched" in some manner. For example, you could have a process where the process design was poorly done. The *Process Owner* did a terrible job of clearly defining what actions needed to be taken in the process and it was also not well organized by failing to assure the correct tools required for the job were called out. While processes that are poorly designed such as this would be expected to fail, we would often find that the process avoided human error failures because it had always been worked by the same technician for the last 15 years. With this kind of experience, the technician already knew everything there was to know about the process. Even though the work instructions were unclear, he already knew what actions to take from his experience in working this process many times previously. Even though the process did not call for him to use certain tools, he already knew the correct tools to use. The real concern became what would happen when this process needed to be run when the experienced technician was out sick or on vacation and some other technician without the same experience has to work the process? What happens when the experienced technician retires? We found many processes exactly like this, that were not presently experiencing any human error failures but were really just a failure waiting to happen when the "crutch" is no longer there to support it.

Many of these poorly designed processes that were being "crutched" had high potential consequences. Our group found this to be extremely troublesome. For years we had only looked into human error process failures, but these "crutched" processes had not been experiencing failures. The only way to identify these types of processes was to just look at large numbers of processes and address them one at a time as they are uncovered.

Our group came to realize that we had many flawed supporting processes being worked every day, and some of these had a potentially high consequence.

There was nothing particularly obvious with these processes and their potential for problems would not be identified unless an analysis was done on each of the processes.

Our group bought completely into the idea that we could use CoBRA to efficiently and accurately analyze processes by focusing on Process Ownership, Process Design (DATOM), and Process Control Strength. Using CoBRA we had the necessary tool to now identify processes with problems, even when there had not been a history of problems. What the group needed to do was just analyze as many processes as possible with the resources we had available to us. Since we knew we had a significant number of processes with problems, the more we could analyze, the more we would identify, and the more problems that could be prevented.

Our manager felt this activity to be so important that he allocated part of our group of engineers exclusively to analyzing processes not presently displaying any failure history. The other engineers in our group would continue to investigate human error process failures as they occurred. The main difference being that when investigating these new failures using a CoBRA method we would also look into any smaller supporting processes to verify they also had no problems. The more processes we analyzed, the more proficient we became at using the CoBRA method to find more and more processes with problems.

This group analyzing processes presently not having

problems worked closely with a team of QC inspectors specially assigned to a program called Process Sampling. These QCs would be sent into randomly chosen work areas and observe ongoing processes, checking specific attributes for compliance. These QC's would document any situation of *Process Workers* not complying with established requirements. When a Process Worker is knowingly taking actions that are not in compliance with the approved process plan, they have improperly taken over ownership of the process. In effect, these QCs were directly identifying for us processes that had Process Ownership issues. Processes that are likely to fail have flaws and weaknesses with Process Ownership, Process Design, or Process Control Strength. Of these three, Process Ownership issues are the hardest to identify because they require someone who is knowledgeable of the process to actually be out witnessing the process as it is being worked. These Process Sampling QCs were extremely useful in the effort to identify processes that had flaws and weakness prior to their actual failing.

The way Shuttle processing operations were organized, was that typically a systems engineer was the individual *Process Owner*. As I said previously, these engineers were very intelligent and highly motivated. You could not find a better group of people to interact with in any company anywhere. These systems engineers in their roles as *Process Owners* were tasked with creating new processes when needed or modifying existing processes as required. The process ownership role kept systems engineers very busy and they were constantly stressed to keep up with the schedule to avoid any launch delays. These process owning engineers understood that, regardless of how much work backlog they had to deal with, NASA also expected them to give a priority to correcting any part of their processes that had already contributed to a human error failure.

By using the CoBRA method, we were finding a number of serious process issues where there were no human error failures yet occurring. We understood that we were identifying process weaknesses, with only a potential for a human error failure, and not an actual failure. The process owning systems engineers were a great group of folks to work with, but asking them to change processes that had not failed was not always well received. These engineers were already stressed over the backlog of things they had to do, and fixing something that had no history of any failures was not given a priority.

Our management understood that it did no good on our part to find a potential problem with a process if we could not get the problem addressed by the already over tasked *Process Owners*. We knew from using the CoBRA method exactly what needed to be changed in the process to eliminate the weakness, but we were not the *Process Owners*, therefore, we did not have the authority to make the changes ourselves.

Our management called the group together and told us that we had to find a way to address in a timely manner these potential process problems we were finding. We all agreed that the process owning systems engineers had a very valid point, and it was unreasonable to expect them to respond to potential process problems with the same urgency that they responded to failures that had already occurred.

Our manager also strongly believed that he did not want our group to be in a position where we had identified a potential for a serious consequence and just accepted that we could not do anything about it. Our group could not tolerate the possibility of someone being hurt or possibly having a serious problem with the Space Shuttle due to a process weakness that we were aware of and were waiting for someone else to correct.

We were fortunate to have in our group a couple of engineers that had previously been Shuttle systems engineers and the *Process Owners* of some very critical processes. These engineers were very familiar with writing process changes. Our manager again broke our group down into another subgroup made up of these previous systems engineers that were knowledgeable with writing process changes. Our manager's new plan was to cease the present process of pointing out a process design flaw or other potential weakness to the *Process Owners* and waiting for them to have an opportunity to make the necessary changes. Our manager's new plan was to take instances where a CoBRA finds a potential process weakness and have our own engineers from our new subgroup write the necessary process design changes themselves. These "ready to go" changes could then be given to the *Process Owners* for their review and signature. This would speed up the implementation of the new process changes by minimizing the workload impact on the actual *Process Owner*.

So now our group had been broken down into three subgroups with all of them using the CoBRA method. We had the subgroup with the original function of investigating human error process failures that had already occurred as well as any smaller supporting processes related to the failed process. We had the second subgroup that was analyzing some of our more serious consequence critical processes, but this analysis had not been kicked off by any failures. This second group was being greatly assisted by the Process Sampling QCs in identifying Process Ownership issues. Finally, we had the third subgroup that had been previous process-owning systems engineers and were addressing potential human error failures by writing the needed process changes themselves, to be reviewed by the actual *Process Owners,* in order to speed up the needed process changes.

You would assume that breaking our group up and taking on two new functions, in addition to our regular function of addressing failures as they occur, would have been a serious impact on our resources. Fortunately, that was not the case at all. When we started using the CoBRA method, the result was that our original function of investigating human error process failures was now being performed much faster. Using CoBRA allowed us more quickly to identify the process flaws and weaknesses that resulted in the human error process failure. CoBRA also told us exactly what corrective actions we needed to implement to prevent a recurrence. Once we started to focus more on the process being worked when the failure occurred, and less on issues surrounding the individual that made the error, our investigations began taking much less time to perform. The resulting increase in investigation efficiency allowed for our manager to move personnel into the two newly created functions without any negative impact on our original major function.

14 - CHANGES BEGIN TO PAY OFF

All of these changes improved our human error failure investigations as well as something we had not anticipated. We began to have a much better relationship with the technicians and their management. Prior to using the CoBRA method, our investigations focused on the individual involved with the human error process failure. As a result, most of our investigation findings came down hard on the technician who was working the process when the error was made. When we began using CoBRA and focusing on the process rather than the individual, our investigations ceased being about finding evidence to prove some technician at fault. When the technicians and their managers saw that our group began taking the position in investigations that we were more concerned about process flaws and weaknesses then trying to blame technicians for making a mistake, they provided much better cooperation in our investigations.

The technicians' managers also appreciated that our group stressed that everyone makes mistakes every single day, so just blaming the technician without fixing the flaw or weakness in the process isn't going to prevent a recurrence of the problem. The managers agreed with us that nobody is perfect, and even the very best technician has the potential to make a mistake at any time. These managers also knew that they usually gave the most complicated tasks to their best people to work. The more complicated a task is, the more likely it is that an error can occur, even with the best people working the task. If human error mishap investigations find that the same technician names keep showing up in problems, it can easily be misunderstood that these individuals are the source of the problems, when in fact, they are the very best technicians. It is often the case that the individuals the managers trust the least, will have the fewest problems because they are never given the most complicated tasks. It is always best to just let local management deal with personnel issues and the investigations need to be strictly about making processes better.

As we stressed this idea of how important Process Ownership, Process Design (DATOM), and Process Control Strength was in the prevention of human error process failures, the technicians and their managers also started working with us more on preventing process problems. Previously, the technicians and their managers had just accepted that some *Process Owners* were very good at building their processes while some other *Process Owners* were not so good. The technicians and their managers just accepted that some processes were going to be poorly organized or somewhat confusing and it was just the technician's job to struggle through these poorly designed processes.

As we worked more closely with the technicians and their managers, we started to get calls from them when they were asked to work a process that had flaws and weaknesses combined with a serious consequence if done incorrectly. Our group gave top priority to these situations because when a technician went to the effort to contact us and say a process had problems, you could bet the process really had some very serious problems. Before calling us, the technicians normally asked the *Process Owners* to improve the process and they only called us if their concern could not be resolved. A flawed process being identified and getting corrected, before there was an actual human error failure, was everybody's ultimate goal.

Something else that worked out well for us was the fact that our group was relocated to a different building and we ended up on the same floor as the Training department instructors. These instructors were a very dedicated group of individuals. We found that they were eager to be informed of any problem observed with any skill they were instructing and certifying the workforce to have accomplished.

Each morning when our group reviewed the documented problems from the day before, we would take special note of any issue possibly associated with any skill taught by the Training department. As soon as it could be determined exactly why the error was made, we could just walk down the hall and talk to the specific instructor related to that skill. The instructors sincerely wanted their classes to properly prepare and certify the workforce and if there was a problem they wanted to quickly address it. We would share with the instructor the information we had and the instructor would then decide how best to modify the class instruction to emphasize the potential for the problem. The instructors would implement what they determined necessary for the very next class coming through.

This was found to be a quick and effective way to address potential human error failure processing problems and it really worked so well because the instructors and their management were professionals who cared about what they were doing. It was purely a coincidence that our group and the Training department ended up being located near each other, but our manager often stated that he wished we had always been located next to the Training organization.

15 - COBRA RISK ASSESSMENTS

Now that our new processes were showing positive results, the group wanted to see if we could develop some additional uses for the CoBRA method. We were successfully using CoBRA in investigations to analyze processes to determine why they failed. We were also using CoBRA to identify processes that were not presently experiencing any problems but were eventually very likely to, due to flaws and weaknesses in the process. The very next use for our new CoBRA method that came to mind was for doing Risk Assessments.

NASA considered risk assessments to be extremely important and they required them throughout Shuttle processing. As stated previously, with a typical risk assessment you need two key pieces of data. You need the realistic potential worst case Consequence of the human error and the Likelihood of the human error and worst case consequence occurring. When it is hardware that is being looked at, risk assessments are a very repeatable, accurate, and useful tool because the Consequence and Likelihood values are data driven and not based on guesses. On the other hand, risk assessments that deal with human error have a much more difficult time. The Consequence is the potential worst case that could result from a human error failure in the process being evaluated. The required Consequence value in a risk assessment of a human error during the process is based on the same basic design information as a hardware risk assessment. The Likelihood value is what is so difficult to determine and where problems arise. You do not have the same Likelihood information with a human error failure as you have when performing a hardware failure risk assessment. With humans, there are no test samples that you can run until failure to base a likelihood value upon. In our situation, no two processes were ever exactly the same and certainly, no two humans were ever exactly the same.

The risk assessment training that was given to our group and the other engineers working on the Space Shuttle program required us to come up with a Likelihood value based on some type of "best guess". Because of this need for guessing, our group did not find these assessments to be repeatable and therefore not reliable either. Risk assessments were used to do some important functions such as prioritizing the order for certain repairs or modifications. We saw a serious concern with the fact that when two separate groups of engineers were given the exact same background information on a problem, one group often arrived at a risk assessment with a high value while the other group might determine a low risk value.

I always liked the simplicity of **Murphy's Law – "Anything that can go wrong will go wrong"**. I have never encountered anyone who disagreed with this simple law. In fact, my experience has always been that whenever the subject of Murphy's Law came up in a group discussion the subject being originally discussed would be set aside and the group would begin exchanging competing stories on how they had been bitten in the past by a situation that seemed so incredibly unlikely to go wrong, but in their situation it surely did. The very basis of how typical risk assessments were being performed was based on just the very opposite of Murphy's Law. The entire premise of the typical risk assessment process was **"Anything that can go wrong will go wrong unless it is unlikely to happen. When something is unlikely to happen, there is no need to worry about it"**. If everyone agrees with Murphy's Law that anything that can go wrong will go wrong, then why are we not using risk assessments that were based on that very same belief? My group always thought that when the consequences are serious, Murphy has the right idea.

What we were seeing was that the engineers had no problem being consistent on the risk assessment Consequence value, but they were all over the place on the Likelihood value. The fact is that the engineers are human, and by nature, some humans are very optimistic while some are very pessimistic. The very optimistic engineers can't see why the error and consequence would ever occur while the most pessimistic engineers believe the error and its consequence will happen the very next time the process is run. What actually happens then is that the engineer who happens to have the most forceful personality will lead the other engineers in the group to agree with his opinion. Why would anyone seriously argue with their co-workers when in reality you are debating who has the best guess on the Likelihood value?

If the most forceful engineer in the group believes the Likelihood value should be low, then the others in the group usually go along and the resulting final risk assessment ends up being determined a low risk. If the most forceful engineer in the group believes that the Likelihood value should be high, then the others in that group go along and the final risk value ends up being a high value. In our opinion, it is only when you can give several groups the same background data and they all come back having determined the same risk assessment final values that you have a risk assessment method that is actually useful.

Since the weak spot in doing a typical risk assessment for human error problems was the determination of the Likelihood value, our group wondered if the CoBRA method could be used to obtain the Likelihood value. We used a 5X5 score card with 25 (5x5=25) being the highest final risk value and a 1 (1X1=1) being the lowest final risk value. First, we looked at the Process Ownership part of the tool and it made sense to us that if there is confusion as to if the *Process Workers* performing the process believe they are allowed to deviate from the process, then there is a high likelihood this process will fail. Therefore, if ownership confusion is identified in the process, then the Likelihood value for the risk assessment would be the highest value of 5. If our group found no Process Ownership confusion, then no Likelihood value would be assigned at this time and we would move on to the Process Design (DATOM) part of CoBRA.

The group would check to see if the process being

evaluated in the risk assessment had all actions clearly laid out with no confusion (Define). We would verify if the process clearly called out all individuals that would be needed for the job and everyone knew the specific actions they would be taking (Assign). We would verify if the process needed anyone to have any particular skills or experience when doing the work, and if so was that clearly made known (Train). We would see if all supporting processes appeared to be working well, the work environment was suitable, and any special tooling was clearly identified (Organize).

Finally, our group would see if the local management for the *Process Workers* performing the process was active in assuring that their personnel performed this task correctly. We would expect to see that some performance data was being tracked to show any evidence of improvement or decline (Monitor).

If the risk assessment evaluation showed any problem with Process Design (DATOM failed), then the risk assessment Likelihood would be assigned the highest value of 5. If the risk assessment showed no Design concerns, then no Likelihood value would be assigned at this point and we would move on Control Strength, the third part of CoBRA.

Our group established early on that the processes most likely to encounter a human error failure would be those that failed Process Ownership, failed Process Design, or that a Control Strength was weak. On the opposite side of the coin, the processes least likely to encounter a human error failure would be those that passed Process Ownership, passed Process Design (DATOM), and were implemented using controls that had a strong Control Strength value. It follows naturally that if you first verify a process passes both Process Ownership and Process Design, then the only remaining variable is the level of Control Strength.

We now had a simple and accurate way of obtaining a repeatable Likelihood value for doing our risk assessments. First, we just analyzed Process Ownership and Process Design (DATOM) and if either fails, then Likelihood value is the highest, a value of 5. If the process being analyzed passes both Process Ownership and Process Design screening, then the risk assessment Likelihood value depends strictly on its Process Control Strength.

Our group had previously identified the 14 controls we commonly used on a daily basis to direct most of the various processes used to prepare a Shuttle for launch:

1. Mechanical Controls / Lockouts
2. Software Controls
3. Written procedures with various combinations of buyoffs required (Tech, QC, NASA QC, Systems Engineer, etc.) verifying the task was properly performed with final product inspection and testing.
4. Written procedures without buyoffs combined with independent final product inspection and testing.
5. Written procedures with buyoffs, but without

independent final product inspection and without testing.
6. Written procedures without buyoffs, without independent final product inspection and without testing.
7. Warning Placards / Maintenance Placards / Operating Instructions Placards
8. Direct On-Site Real-time Instructions
9. Training with Competency Certified
10. Bulletins
11. Tailgate Meetings
12. Non-Certified Training
13. Workforce Standard Procedures or Desk Instructions
14. Posters, Quality Slogans

These 14 controls that we commonly used had all been ranked on relative strength. We then took these 14 controls and assigned each a value of 1, 2, 3, 4, or 5. After much discussion, our group assigned the very strongest controls, 1 through 3, a Likelihood value of 1. We assigned controls 4 and 5 a Likelihood value of 2. Control numbers 6-9 were assigned a Likelihood value of 3. Controls 10–13 were assigned a Likelihood value of 4. The weakest control, number 14, was assigned the Likelihood value of 5. What is most important here is that we all arrived at a final agreement on which controls were assigned to each Likelihood value. The controls and their agreed upon corresponding Likelihood values were then locked down in a list and copies were provided to everyone in the group.

Everyone in our group began performing risk assessments using the same method. First, we had to determine if there were any problems with Ownership or Process Design (DATOM) and if there were problems, the Likelihood value was assigned a 5. If there were no problems with Process Ownership or Process Design (DATOM), then we merely obtained the Likelihood values based on which control was being used in the process and its assigned corresponding value from our list. Since there already was consistency in the Consequence values, the entire group was now all arriving at the same exact Consequence and Likelihood values.

For example, if we were doing a Risk Assessment on a process, we would first assure no problems with Process Ownership or Process Design (DATOM). We would then see how critical the engineers responsible for the process felt the process to be. If these engineers determined that a human error in the process could result in a death or serious injury, then there would be no doubt that the Consequence value would be a 5. To get the needed Likelihood value, we would just look at what control was being used to manage the process. If we saw the process was being controlled with written work steps, with technicians buying off each step as performed, with independent QC verifications from our QCs and NASA QCs, combined with final product inspection and testing, we would then look at our list and see what Likelihood value this particular control was assigned. We find written work steps, tech buy-offs, independent QC verifications, and final inspection and testing, is a high strength control with a previously agreed upon corresponding Likelihood value of 1. Our resulting risk assessment would then be a Consequence 5 multiplied by a Likelihood 1 giving an overall Risk Value of 5.

We believed strongly that risk assessments performed using this three part CoBRA method were a major improvement over how we were previously doing risk assessments. Doing risk assessments using a CoBRA was repeatable and based on a rationale that everyone could readily understand. We would regularly have to present to management our suggested corrective actions at the end of an investigation. These presentations would include a required risk assessment of what the risk value with our suggested corrective actions implemented compared to what the risk value would be without the suggested corrections. When my group, using a CoBRA Risk Assessment, identified a high risk (Red) as a result of some process flaws we would point this out to our NASA counterparts. What would often happen was our NASA counterparts would then use one of their own types of risk assessment methods. The NASA risk assessment methods usually determined a lower final risk value than a CoBRA, if there was no history of any previous failures in the process. Even though our NASA counterparts often arrived at lower final risk values, they would nearly always agree that our risk assessment had identified some process flaws and weaknesses that needed to be corrected.

Our group was never upset whenever this occurred because we never believed that the purpose of a risk assessment was to arrive at a certain color or risk value, but rather as a tool for identifying flaws and weaknesses needing to be corrected. We understood that the CoBRA risk assessment was often more pessimistic than other types of risk assessments because the CoBRA gave no credit for any past history of being problem free. The fact that our NASA counterparts agreed that our CoBRA risk assessment had identified flaws and weaknesses needing to be fixed, was really the issue that was important to us.

16 - METRICS INDICATE SUCCESS

Within a couple of years, our group had significantly changed how we were doing business. We had now developed, and were using, our own investigation method based on Process Ownership, Process Design (DATOM), and Process Control Strength that we called a CoBRA, which was short for Control Based Risk Analysis. We had also changed the groups' organization by having it broken down into three separate teams, with each performing different functions. One team continued to look into the daily human error processing failures as they occurred as well as looking for any additional problems in any supporting processes. This first group also immediately investigated all failures of strong controls. The second team moved into more preventative type actions by evaluating highly critical processes and their key supporting processes for flaws and weaknesses. The third team assisted process owning engineers by drafting the necessary process changes themselves when weaknesses in processes were identified. This allowed for weak or flawed processes to be fixed as soon as possible with less impact to the already busy *Process Owners*.

Our local management was very happy with what we were now doing, but more importantly, our NASA counterparts responsible for monitoring our group were also pleased. Our NASA counterparts were a great group of engineers to work with, but they had a serious job to do and they expected to see results from our actions. If our NASA counterparts ever failed to agree at any time with what we were doing, they could have easily forced us to change our methods. Fortunately for us, our NASA counterparts agreed with the changes our group had undertaken and particularly liked the emphasis being made into the area of preventative actions.

NASA could have shut down what we were doing at any time, and our own company also could have put a quick end to our effort. My group was in the Quality Directorate, and the changes we were making often had an impact on the Engineering Directorate. My group was very fortunate that there were a few engineers within the Engineering Directorate that held staff positions that saw the benefit to what we were doing. The opinions of these staff engineers were trusted by upper management in the Engineering Directorate and their support for many our process changes was critical for getting these changes implemented. These staff engineers, who supported what we were doing, realized the need for developing tools that could identify process weaknesses before there was an actual mishap and they saw potential in what my group was doing. I cannot stress enough just how easily everything my group was doing could have been shut down without the support of these staff engineers. I am sure they have all gone on to other companies and taken with them many of the basic ideas of CoBRA and are using these concepts with their new companies.

We had the support of our NASA counterparts and key individuals within the engineering organization. Everyone concurred with the logic behind a CoBRA and the way it was being used. The only thing that now remained was to monitor the ongoing data being collected to see if any reductions in human error process failures could be observed when comparing how we did business in the past to what we were now doing. The obtaining of accurate data on the use of CoBRA and any benefit it may, or may not be providing, was not an easy task and it was the next hurdle our group had to face.

My company had the contract to process the Space Shuttle for launch and that included all of the ground supporting equipment as well. NASA required the detailed documentation of all problems encountered during processing. Since we were documenting problems occurring with both flight hardware and ground support equipment, the documentation of thousands of problems in a month was not uncommon. Fortunately, the vast majority of problems were hardware related and not workforce issues.

Our group focused almost entirely on workforce human error problems occurring during Shuttle processing. There were several other engineering groups that dealt with hardware and design problems while our group only focused on workforce problems. Early in our group's existence, a manager had us create our own database for all documented problems that our group addressed. While there were several different engineering groups creating different databases for tracking various Shuttle processing problems relating to their individual groups' activities, our database collected a very specific type of information.

To repeat what was previously stated, every morning, our entire group sat down with our NASA counterparts and coded all documented Shuttle processing problems from the previous day. We broke down every problem into one of two categories that we referred to as "Preventable" and "Non-Preventable". The Non-Preventable problems were those that our workforce had no real control over, such as flight damage, hardware wearing out, or weather related damage. The Preventable group of problems, on the other hand, were the problems that could have been caused by workforce human error and also damage caused by personnel as a result of physically working in and around the hardware.

We then took the Preventable group of documented problems and broke them down further into "Preventable Collateral" and "Preventable Processing". The Preventable Collateral problems could be best described as being caused by humans performing normal workforce activities around hardware that was easily damaged. We found most of these documented problems to just be the result of our technicians having to perform tasks in tight spaces and difficult work areas. Tools sometimes slipped and damaged things. Sometimes things got stepped on and crushed. Maybe switches got bumped and ended up in the wrong configuration. The Preventable Collateral human error failures were the type that was best addressed by telling the workforce that they needed to be more careful, while in reality, we knew they were probably already doing the best they could in the difficult conditions they were working in. It was just hard to keep from damaging things when you are climbing in and out and working in confined tight areas with nearly everything around you susceptible to being damaged in one way or another.

We found the best thing about tracking Preventable Collateral human errors is that they are a great indicator of work volume. Sometimes there would be three Shuttles being processed at the same time, and during other times there may only be one actually being processed with the others in orbit or at the pad awaiting launch. We observed that in the months when work volume increased, there was a corresponding increase in Preventable Collateral documented problems. Likewise, when work volume decreased, so did the Preventable Collateral problems. We found that work volume correlated directly with the number of Preventable Collaterals. There was nothing surprising about this observation. Preventable Collateral problems result from people doing work, so it made sense to us that the more work people are performing, the more instances of inadvertent worker damage that will be documented.

The other group of documented preventable human errors was the Preventable Processing problems. These were the errors that my group focused on, the ones we felt we should be able to prevent. These were the errors that the changes our group implemented would have had an impact on. The Preventable Processing errors were the mistakes the workforce made involving a process failure caused by some type of incorrect decision making and not because they were performing work in a difficult environment. If a worker slips and inadvertently bumps a switch, putting it into the wrong position, then that error was a Preventable Collateral. If that same worker incorrectly read a confusing procedure and flipped the wrong switch, then that error was a Preventable Processing error. While the result was the same, a switch in the wrong position, the misreading of the confusing procedure was the type of error our group felt we had the better opportunity to prevent.

We found that by tracking the ratio of Preventable Processing errors per the number of Preventable Collateral errors, we could get a fairly accurate measurement value for the amount of human error failures per a standard volume of work being performed. This was, to us, the best way we could come up with to determine success or failure of the changes we had put in place using CoBRA.

The year 2006 was the year we began to implement CoBRA as our standard way of doing things. Using 2006 as our starting baseline, we observed that our workforce was experiencing an average of 162 Preventable Processing errors per month per 36 Preventable Collateral errors. This created a monthly average ratio value of 4.5/1 Preventable Processing errors per each Preventable Collateral error for the year of 2006. In 2007, all of our new methods had been up and running for nearly an entire year and during 2007, our workforce per month averaged 140 Preventable Processing errors per 40 Preventable Collateral errors. This created a monthly average ratio value of 3.5/1 Preventable Processing errors per each

Preventable Collateral error for the year. By 2008, our new methods had been in place for two years and the workforce averaged 65 Preventable Processing errors per month compared to 37 Preventable Collateral error per month. This created a monthly average ratio value of 1.8/1 Preventable Processing errors per each Preventable Collateral error for that year. In 2009, the new methods had been in place for 3 full years and the workforce averaged 58 Preventable Processing errors per month compared to 35 Preventable Collateral errors for the ratio of 1.7/1. In 2010, the new methods had been in place for 4 years and the workforce was averaging 43 Preventable Processing errors per month compared to 31 Preventable Collateral errors. This created a monthly average ratio value of 1.4/1 Preventable Processing errors per each Preventable Collateral error for the year of 2010.

In 2011, NASA shut down the Space Shuttle Program which resulted in our company and our group being dissolved. The data we had up to this point indicated to us that the new methods we had employed had been very successful and we went from a monthly average of 4.5/1 to 1.4/1 Preventable Processing errors to Preventable Collateral errors, a reduction of nearly 70% in five years to what had already been a mature program when we started.

We were seeing such positive results and we wanted to share the ideas that we believed contributed greatly to these results. Unfortunately, the timing was horrible. Our company was going to disappear when the Shuttle program ended and the attention of our management was focused clearly elsewhere. Our upper management was preoccupied with having to lay off thousands of employees and they were trying to make the process as painless as possible for the workforce. Normally, NASA would have jumped on this kind of data if it had come at another time in the Shuttle program. Earlier in the Space Shuttle program, this type of positive data would have resulted in a trip to NASA headquarters to present it to NASA upper management. At this point in the Shuttle Program, it was just an improvement to a program that was going away. Sadly, the positive data and the things we learned meant little to anyone due to the bad timing.

17 - RUN THE TWELVE ITEM CHECKLIST

As the Shuttle Program came to an end, everyone understood that our group was going to be dissolved and we were all going to be heading off in different directions. This was a very sad time, but also one of new opportunities. One of the things the group discussed was what exactly would we bring to a new company based on what we learned during our time with the Shuttle program.

Something we all agreed on was that when a serious mishap occurs due to human error, it is not the result of any kind of unavoidable bad luck, but rather the natural outcome to be expected when processes are poorly enforced, poorly designed, or poorly implemented. The quality of an organization's processes is what determines if that organization is going to be a victim of "bad luck" or not.

I never liked the way the "Swiss Cheese Diagram" was used when describing what happened in mishap investigation presentations to upper management. It seemed like every time a serious mishap occurred and the local manager responsible for these failed processes had to explain to upper management what happened, they would fall back on the "Swiss Cheese Diagram". A slide would be flashed up on the screen showing four or five slices of Swiss cheese. Each slice representing one of the four or five supporting processes that failed in some manner, ultimately resulting in the serious mishap currently being investigated. This diagram would always show nearly solid slices of Swiss cheese with only a few small holes representing weaknesses that existed in the process. Finally, there would be an arrow representing the mishap going through all the slices where these very small holes just happened to line up by some rare instance of bad luck and allowed the mishap to get through.

These Swiss Cheese diagrams were believed to be the perfect tool for explaining how these managers were actually excellent managers, but they just happened to be the victim of very bad luck in this particular incident. The managers responsible for the failed processes would then try to explain that there was no way that anyone could have foreseen that this series of unfortunate failures in supporting processes could have ever taken place all at the same time.

In reality, what these Swiss Cheese diagrams were actually showing was that this incident was the result of having numerous unreliable supporting processes. Holes lining up is not the problem; having those holes there to begin with is the problem. For any process to be successful, you have to be able to count on all supporting processes. I would see over and over again how if a simple CoBRA risk assessment had been performed on these supporting processes prior to the mishap; it would have quickly revealed in advance the flaws and weaknesses that clearly existed.

Depending on the size of a company, there can be hundreds of processes, or even hundreds of thousands of processes, going on daily. Everyone in the workforce, from the time they get to work until they leave, are involved with performing different types of processes. With some processes, a human error will have minor consequences, but other processes can have very serious consequences if a human error should occur. Our group considered a human error to be any situation where the actions taken by the personnel working the process deviated from what the *Process Owner* intended. It would be obvious to everyone that an error in the process of filling a pressure vessel can have a serious consequence if it is over-pressurized, but the process of going down a flight of stairs at your office can also have a serious consequence if errors are made. A worker is just as dead who has fallen down a flight of stairs as a worker who is killed by a pressure vessel exploding.

Our group agreed that one of the most important things any organization could do, was performing risk assessments on all processes that could possibly have a potential for a serious consequence if a human error occurred. Risk assessments are the common language that allows Quality, Operations, Safety, Engineering, and Management to all equally identify and understand a flawed process and determine if it needs to be immediately addressed or if resources would be better spent elsewhere.

Whenever we had a serious consequence human error investigation, there were always two elements that seemed to be present:

#1- The local workgroups always had some previous

knowledge of the flaws in their own processes, they just didn't see them as being that serious or in need of immediate attention. These flaws were never major or flagrant violations; they always were just weaknesses that never really caused problems before. Then something happens that causes these processes previously considered to be unimportant to become very important. For example, failing to replace burned out overhead light bulbs in a work area is not a major problem until someone misreads a switch nameplate due to poor lighting and activates the wrong equipment during a critical process. Supporting processes that appear to be of low importance often have the potential to be very important.

#2 - In each serious consequence incident, management seemed to be surprised and disappointed because they were under the impression that major critical processes would be supported by other processes that were as robust and well maintained as the more visible critical processes. Smaller supporting processes often get overlooked, and it is not until they negatively impact a more critical process that they happen to be supporting that their flaws and weaknesses are realized.

The most troubling thing about these serious consequence mishap investigations was that the problems in these failed supporting processes were not difficult in hindsight to recognize, and management never had an issue with providing the resources needed to correct the problems once they were identified. These human error mishaps could have been easily avoided had the problems with the supporting processes been recognized and addressed prior to the mishap. The core of the problem was that there was no mechanism that called for the analyzing of processes that were not presently having problems. We had all kinds of mishap review boards, investigation teams, and dozens of other mechanisms for analyzing processes after a problem occurred, but what was truly needed was a way to continually analyze all processes to identify flaws and weaknesses before mishaps took place.

Our group had good results in significantly reducing the negative consequences that were occurring as a result of human error, but we were 10 engineers with additional assistance coming from a team of Process Sampling QCs. Few companies would ever have those kind of resources available to strictly dedicate to investigating and preventing workforce human error issues. Something we realized at the end of all of this was that when using CoBRA, you do not need a separate group of individuals for addressing human error concerns. The local workgroups themselves can be utilized to identify and correct their own processes. Nobody understands these supporting processes better than the local workgroups that actually perform these processes. These local workgroups know when their processes are not being worked exactly as called for by the *Process Owner*. In order to get ahead of human error problems, a company has to provide the local workgroups with the means and incentive to identify and address these process weaknesses before there is a human error failure. A risk assessment is a perfect tool for enabling local workgroups to evaluate their own processes and recognize existing weaknesses before there is a problem. A risk assessment is also a great means of documenting weaknesses and potential problem situations, and thus preventing them from "falling through the cracks" and being overlooked.

There is no better means for dealing with potential human error than having local workgroups continually evaluating their own processes by doing their own risk assessments. The use of the CoBRA concept and methods makes it simple enough for local workgroups to do their own risk assessments on their own processes and get results that are accurate and repeatable. When my group did a CoBRA risk assessment, we were never the experts on the process being assessed and therefore, it would often take several hours or longer to get a final risk value. When a local workgroup does a risk assessment on their own process, they are already experts on the process so the CoBRA Risk Assessment can be done much faster, typically in less than an hour. This much faster version of the CoBRA Risk Assessment due to being done by the local workgroup experts is where the term CoBRA 30 Minute Risk Assessment comes from. The basic idea is still the CoBRA risk assessment, but it is instead called the CoBRA 30 Minute Risk Assessment due to taking much less time to perform when done by the local workgroup experts doing their own processes.

When a workgroup performs a risk assessment on their own processes, the determined final risk value is far more accurate and useful than when an outside group attempts to do it. If a final risk assessment value is determined to be high, then local management understands they must look into this process much more closely and see if there is any action that they can take at their own level to reduce the high risk. If the actions required are beyond the ability of local management, the risk assessment is a great tool to explain to upper management exactly what the concern is and why the assistance of upper management may be warranted. A properly done risk assessment is easy to understand and difficult to argue against. A risk assessment is the best tool available for understanding a potential problem and conveying the concern to others.

An organization could see significant benefit by expanding the use of risk assessments beyond having a few specially trained individuals evaluating a limited number of critical processes. All workgroups can easily and quickly perform a CoBRA 30 Minute Risk Assessment on their own processes when provided with the following two pieces of information by management:

1 - A basic company-wide list of a range of possible outcomes that might result from any human error problem. These possible Consequences would be broken down into five different ranges and a 1-5 corresponding value for each range of Consequence. For example, "death or serious injury" is assigned a value of 5, while a cost impact to company of "$100 or less" gets assigned a Consequence value of 1.

2 - A basic company-wide list of the common controls the company uses and their corresponding Likelihood

values that company management has previously agreed upon. For example, "detailed written work instructions with a QC witness" is assigned a Likelihood value of 1, while a "Work Safe" wall poster being used as the control gets assigned a Likelihood value of 5. There also needs to be a clear description defining each control and it needs to be made very clear how each control should function. While the list may state "written work steps with QC witness" there also needs to be a clear definition for what a "QC witness" actually involves. For example, the QC shall be thoroughly trained and experienced in the work being observed and the QC must witness and independently concur with each action being taken. A QC that only later, after the fact, confirms in some manner that the work had been performed would not, in this case, meet the description of an acceptable "QC witness" control.

The final risk assessment value is determined by first verifying no issues with Process Ownership or Process Design (DATOM) and then multiplying the Consequence number by company assigned corresponding control strength Likelihood number. With the final risk value known, company management might also want to provide direction on what actions are warranted by which risk assessment scores. For example, for risk assessment values of 10 and below, the company may direct that no actions are required by workgroups. For scores 11 and above, local workgroups might be directed to explore potential actions that could lower the value to less than 11. For risk assessment values of 15 or above, workgroups might be directed to also notify upper management of the high-risk process.

It is not complicated to perform a CoBRA 30 Minute Risk Assessment at the workgroup level. The CoBRA 30 Minute Risk Assessment simply provides the local workgroup with an easy method where they can identify if one of their processes has problems with Process Ownership, Process Design (DATOM), or Process Control Strength. Accurate and repeatable CoBRA 30 Minute Risk Assessments values can be determined with just the most fundamental knowledge of how the CoBRA 30 Minute Risk Assessment works along with the management provided basic Consequence and Likelihood range lists

Being simple and quick makes a CoBRA 30 Minute Risk Assessment a very useful tool for local workgroups to generate Risk Assessments on their own processes. A workgroup only needs to ask a few basic questions to get started;

1. Can the group think of any type of mistake that anyone in this workgroup could make that might result in an injury to themselves or anyone else in this workgroup? Any "close calls" or "near misses" that anyone may have observed or heard about?

2. Can the group think of any type of mistake that anyone outside this group could make that might result in an injury to anyone in this workgroup?

3. Can the workgroup think of a mistake that anyone in this workgroup, or anyone outside this workgroup, could make that might have a negative impact on the

product or services this workgroup provides to the company?

Once the workgroup has identified any potential mistakes that could meet any of these three criteria, then the question that follows is "What processes do we have in place that are intended to assure that we are not negatively impacted by any of these potential mistakes?" Each separate process that is identified needs to be looked at and evaluated. To do a good CoBRA 30 Minute Risk Assessment the local workgroup only has to complete the following twelve item checklist:

1. (Ownership) Who exactly owns this process being analyzed and do the workgroup members understand that, as *Process Workers*, they cannot vary from what the *Process Owner* describes as necessary to be done. Are there any situations or exceptions when the *Process Owner's* desired plan might not be followed? Has the *Process Owner* been made aware of these situations and exceptions and has he given approval? (Any issues makes the Likelihood value a 5)

2. (Define) - Is the process well defined and clear as to exactly what actions need to be taken with nothing in the process being confusing or overly complicated? (Any issues makes the Likelihood value a 5)

3. (Assign) - Is the process clear as to who exactly is supposed to be doing each particular action in the process? If for some reason the designated person tasked with the actions gets called away, leaves for a

break, goes out sick, leaves on vacation, or has a family emergency, is it clear who takes over and are they clearly called out as an equally suitable replacement? (Any issues makes the Likelihood value a 5)

4. (Train) - Does the process require any particular skills that are not directly called out in the process as being required? If the time required to complete the task is important, do the *Process Workers* have the skills to work fast enough to complete the task in the time allowed? (Any issues makes the Likelihood value a 5)

5. (Organize) - Does the process do an adequate job of assuring a suitable work environment and proper tool availability? Is this process vulnerable in any way to the actions of others and are there supporting processes provided by others that could have a negative effect on our processes if the supporting processes have a failure? Do you have total confidence in these supporting processes? (Any issues makes the Likelihood value a 5)

6. (Monitor) - Is local management showing enough oversight of this process that if any weaknesses crept into the process, it would be quickly apparent to local management that there was a problem forming. (Any issues makes the Likelihood value a 5)

7. (Identify Consequence) - What does the workgroup believe to be the worst case scenario human error that might realistically occur during this process?

8. (Consequence Ranking) - From the list provided by management, what is the matching Consequence value for this worst case scenario?

9. (Identify Controls) - What means is controlling or providing directions for this process to assure that it gets worked exactly as the *Process Owner* wanted?

10. (Control Compliance) - From the Likelihood list provided by management on controls, does the control in this process comply with the provided description of how this control should be implemented?

11. (Control Ranking) - From the Likelihood list provided by management on controls, what is the matching Likelihood value for the control being used in this process?

12. (Final Risk Value) - When the Consequence value and the Likelihood value are multiplied together, what is the final risk assessment value?

As an example of using the twelve item checklist in a CoBRA 30 Minute Risk Assessment lets look at the process of assuring forklift drivers do not exceed safe speeds in a warehouse:

Question #1 (Ownership) – Who is the person that

owns this process and gets to set the speed limit for forklift drivers in this warehouse? Is the sign saying "3 MPH Max Speed" taken seriously? Is the required training class that told the drivers not to exceed 3MPH taken seriously? Do the drivers feel that they get to determine the speed they think appropriate? Are there any situations or conditions when it would be acceptable to exceed the 3MPH speed limit? Has the *Process Owner* been made aware of these situations where the 3MPH limit can be exceeded and has he given approval?

Question #2 (Define) – Is there any confusion as to what the "3 MPH Max Speed" means? Could the limit being described be misunderstood?

Question #3 (Assign) – Is there any confusion about who this speed limit sign is intended for? Does every forklift driver understand this sign is meant for them? Do the more senior and more experienced drivers feel that the signs only apply to the newer drivers? Do the drivers feel that whenever there is a "rush order" they can ignore the speed signs in those situations? If the normal drivers were out sick or on vacation, would their replacements equally understand?

Question #4 (Train) – Are there any skills required to driving a forklift and keeping it under this 3 MPH speed limit? Could a person who has never driven a forklift before, just get behind the wheel for the first time and keep it under 3 MPH without difficulty? Do the drivers have the skills to complete their required volume of work while driving within the allowed speeds?

Question #5 (Organize) – How easy is this sign to read? Is it high enough, with large enough lettering for a driver to notice? Is lighting in the warehouse well maintained? How far apart are the signs spaced? Do the forklifts even have a speedometer or any other device to alert a driver if they are exceeding 3 MPH?

Question #6 (Monitor) – If it is a common practice to just ignore this speed limit sign and instead choose to go a faster speed, will anyone in management be aware of this ongoing tendency of failing to comply? Is data such as the number of accidents, number of excessive speed warnings, or number of disciplinary letters issued, being monitored by anyone?

Question #7 (Identify Consequence) – If a forklift driver exceeds the 3 MPH speed limit called for on the signs, what is a potential worst case consequence? Is a wreck, damaged critical hardware, or serious injury/death, a realistic possibility?

Question #8 (Consequence Ranking) – From the company supplied list, what is the matching Consequence value that would go with this identified realistic worst case outcome.

Question #9 (Identify Controls) – What control or combination of controls is being used to assure that the forklift drivers are aware of the 3 MPH maximum speed which the *Process Owner* does not want the drivers to exceed?

NOTE – Controls identified in this example might

include: "3 MPH Maximum Speed" signs throughout the warehouse, mandatory training classes for forklift drivers with a requirement to pass a test upon completion of the class, and safety inspectors writing up drivers who violate the 3 MPH limit.

Question #10 (Control Compliance) – Do the numerous "3 MPH Maximum Speed" signs, the requirement for special training with a test, and periodic safety inspector visits comply with the description on the company supplied basic list of general controls such as "Work Area Instruction Placards" and "Certified Training"?

Question #11 (Control Ranking) – From the company supplied Control Strength/Likelihood value matching list, what number values do these controls match up with?

Question #12 (Final Risk Value) – By multiplying the Likelihood value by the Consequence value what final CoBRA 30 Minute Risk Assessment value is determined?

What if the controls were found to have problems? What if the signs were too small and poorly lit? What if the signs, instead of giving a definite value only said: "Do Not Exceed Safe Speed"? What if some of the drivers were still on the job, but on "probation" because they never passed the final testing part of the training. There are many situations that could be uncovered during the evaluation that might have to be considered before the full credit value of the company supplied Likelihood/Control Strength list can be applied. What would be a very strong control when correctly implemented might actually end up being a very weak control when its flaws are considered.

Strong controls with flaws can be more dangerous to a process than not having strong controls there at all. Just because processes such as QC verifications, software overrides, final inspection/testing, or switch locks are normally very strong controls, they can also lead to a false sense of security if they are not properly implemented and maintained.

Having QC on a task, performing a witness of actions being taken, is only effective as a strong control if the QC is completely independent. A QC cannot assist in an action he is supposed to independently witness. A QC cannot become part of the team performing the actions by taking on a role in the process, such as reading out the steps in a procedure in order for the technician to work them more quickly. The QC and the *Process Worker* must be equally skilled on the actions being taken and the QC is not there to resolve any confusion that may occur when working the process. The QC and the *Process Worker* should never be in a discussion trying to resolve what actions the *Process Owner* desired to be performed. If either the QC or the *Process Worker* has any questions regarding any required action, only the *Process Owner* can clarify that confusion. A human error occurring in a process with QC verification as a control is the type of strong control failure that should always be investigated.

Software is also a great control and can be one of the

strongest available, but it is only as good as the people who wrote it and tested it. If it has not been thoroughly tested, all you know is that it will work in the limited situations it was tested against and nothing more. Software that has not been thoroughly tested is especially dangerous because the people using the software will trust it like it has been tested in all situations, but in reality, they never know when they might be in an untested situation.

Like software, final inspections/testing is only as good as the level of inspection or testing that is performed. If a valve is manufactured and the final inspection/testing only requires that the valve opens and closes, but does not involve a leak test, it cannot be considered to be that strong of a control for assuring the valve was correctly assembled. A mechanical switch lockout can also be a very strong control, but if for the sake of convenience nearly everyone has a key, then what you have is not a strong control at all. When a strong control is flawed it is more dangerous than a weak control because you expect more from it, and when it fails you are less likely to be prepared to deal with it.

Here is another example of a CoBRA Risk Assessment (using the twelve checklist questions) of the type that might possibly be encountered during Shuttle processing:

There are many pieces of equipment that only need to fly on the Shuttle based on the specific requirements of the next upcoming mission. For example, experiment packages, tooling for repair missions, or spacesuits if a mission requires astronauts to perform outside activities. These items are installed on the Shuttle at different points during processing and are often one-of-a-kind and extremely expensive. They are stored in special areas that are highly controlled and manned by logistics storekeepers. When these mission specific items are needed for testing, fit checking, or installation, the storekeepers are notified and the items are removed from shelves and provided to the personnel that will perform the required operations. Originally, each of the special storage areas always had two storekeepers. Several years ago, their management made a cost-reducing decision to only have a single storekeeper to support each area on weekends and second shift.

A risk assessment would usually begin by asking these storekeepers as a group if they can think of any situation that might warrant evaluation, such as a near-miss incident or any kind of confusion encountered when working their normal tasks. One of the storekeepers points out that they could possibly get hurt while lifting some of the heavier pieces of mission equipment by themselves when working alone on weekends and second shift. The storekeepers point out that even though some of these boxes are clearly marked "Two-Person Lift" and also have a picture of two people lifting a container printed on the boxes, the storekeepers are having to lift the boxes by themselves due to being the only person working at those times. The group decides to run a CoBRA Risk Assessment on this situation using the twelve question CoBRA checklist:

1. (Ownership) Who exactly owns this process being analyzed and do the workgroup members understand that, as *Process Workers*, they cannot vary from what the *Process Owner* describes as necessary to be done. Are there any situations or exceptions when the *Process Owner's* desired plan might not be followed? Has the *Process Owner* been made aware of these situations and exceptions and has he given approval? **(Any issues makes the Likelihood value a 5)**

The storekeepers agree that in this situation it is intended for them to be the *Process Workers* and the *Process Owner* is their manager since it is the manager who decided that one person is all that is needed on weekends and second shift. It is also the manager who also tells them they are responsible for following all handling instructions provided on the storage boxes. In this situation, the *Process Owner* has created confusion regarding what actions are required by the *Process Workers*. Their manager (*Process Owner*) requires them to comply with handling directions such as "Two-Person Lift", but the storekeepers are confused by not having a second storekeeper available. Due to this confusion, they have just been ignoring the "Two-Person Lift" handling directions. The group admits that they have all been lifting these "Two-Person Lift" containers by themselves and this has gone on for years. The storekeepers have effectively taken over ownership of this process. If the storekeepers had initially asked their manager (*Process Owner*) for clarification when this confusion first occurred and resolved the confusion at that time, this would not be an

Ownership issue. Ownership - Fail

2. (Define) - Is the process well defined and clear as to exactly what actions need to be taken with nothing confusing in the process or overly complicated? (Any issues makes the Likelihood value a 5)

The storekeepers agree that while the "Two-Person Lift" handling instructions are clear on the box and even include a picture of two individuals lifting a container, there is confusion created by having only one storekeeper working on weekends and second shift with no additional instructions provided by management for this situation. Define - Fail

3. (Assign) - Is the process clear as to who exactly is supposed to be doing each particular action in the process? If for some reason the designated person tasked with the actions gets called away, leaves for a break, goes out sick, leaves on vacation, or has a family emergency, is it clear who takes over and are they clearly called out as an equally suitable replacement? (Any issues makes the Likelihood value a 5)

The storekeepers all agree that two individuals are required to comply with handling instructions. On weekends and second shift, when only one storekeeper is working, it is not clear to them who the second individual is supposed to be. Assign – Fail

4. (Train) – Does the process require any particular skills that are not directly called out in

the process as being required? If the time required to complete the task is important, do the *Process Workers* have the skills to work fast enough to complete the task in the time allowed? (Any issues makes the Likelihood value a 5)

The group of storekeepers agrees that there is no problem with any additional skills required that they do not already have. Train - Pass

5. (Organize) - Does the process do an adequate job of assuring a suitable work environment and proper tool availability? Is this process vulnerable in any way to the actions of others and are there supporting processes provided by others that could have a negative effect on our processes if the supporting processes have a failure? Do you have total confidence in these supporting processes? (Any issues makes the Likelihood value a 5)

The group of storekeepers concurs that there are no problems with the work environment, tooling, or support processes. Organize - Pass

6. (Monitor) - Is local management showing enough oversight of this process that if any weaknesses crept into the process, it would be quickly apparent to local management that there was a problem forming. (Any issues makes the Likelihood value a 5)

This confusing situation went on for years without being identified by management. The fact that clear handling instructions were not being complied with should have been realized sooner by the local management. Monitor – Fail

7. (Identify Consequence) – What does the workgroup believe to be the worst case scenario human error that might realistically occur during this process?

The group agreed that when one storekeeper is performing lifts that should have been done by two individuals, there is a potential for injury. The likely injury could result in lost time, but not death or disability. This equipment is critical to mission success and extremely expensive. Some pieces are not replaceable. Extensive testing would prevent any damaged equipment from putting Shuttle and crew in jeopardy.

8. (Consequence Ranking) - From the list provided by management, what is the matching Consequence value for this worst case scenario?
For Shuttle processing damage to highly expensive equipment, threat to mission success, or serious injury (not life threatening) to be Consequence Value - 4

9. (Identify Controls) - What means is controlling or providing directions for this process to assure that it gets worked exactly as the *Process Owner* wanted?

The group concurred that the "Two-Person Lift" handling instructions are clearly printed on the box along with a picture of two individuals lifting a container. The job was intended to have a second storekeeper helping with the lift. For the storekeepers this control would have been considered a "Warning Placards / Maintenance Placards / Operating Instructions Placards" but with the addition of a second independent individual verifying the actions. This would have been a very strong control since it more than adequately addresses the potential for humans to forget things, become confused, or become distracted.

10. (Control Compliance) - From the Likelihood list provided by management on controls, does the control in this process comply with the provided description of how this control should be implemented?

The group agreed that in this situation, without the second independent storekeeper, this control would not have been working as intended. Control – Not As Described

11. (Control Ranking) - From the Likelihood list provided by management on controls, what is the matching Likelihood value for the control being used in this process?

The group determined that the control ranking with the second storekeeper would have been a very good value at 2.

12. (Final Risk Value) - When the Consequence value and the Likelihood value are multiplied

together, what is the final risk assessment value?

The group of storekeepers determined that if there had not been the confusion as a result of having only one storekeeper working weekends and second shift, the final risk assessment value would have been an acceptable 8. This value would have been the result of a Consequence value of 4 multiplied by a Likelihood value of 2. When the CoBRA Risk Assessment twelve questions are asked it is found that the questions on Ownership, Define, Assign, and Monitor all fail and therefore a Likelihood value of 5 is assigned. Any failure or concerns with Ownership, Define, Assign, Train, Organize, or Monitor will always result in a maximum Likelihood value of 5. The CoBRA Risk Assessment final value is a Consequence of 4 multiplied by a Likelihood of 5 for a high-risk final value of 20.

This is a problem that initially appears to be easily fixed by addressing the failures with Define, Assign, and Monitor. By putting a clear process in place regarding what to do when two storekeepers are required, combined with having a means for local management to check on this process more closely in the future, this particular high-risk situation would seem to be adequately corrected. With a CoBRA Risk Assessment, the high-risk situation is only partially addressed by assuring two storekeepers perform these lifts in the future. The high-risk situation remains until all of the problems with this process are corrected. That means the final issue with Ownership also has to be addressed before the high risk can be finally lowered.

The fact that all of the storekeepers felt it was appropriate to ignore the handling directions and take over ownership of this process when there was confusion is very serious. Considering that this went on for years without anyone bringing it to management's attention demonstrates that this way of thinking is clearly not limited to only this particular situation. Occasionally a *Process Owner* may fail to identify and address each and every situation in a process where some confusion may exist. When *Process Workers* feel that rather than getting the confusion resolved, they can instead take over ownership of the process, then you can no longer have confidence in any of your processes. Fixing the problem with Define, Assign, and Monitor will keep this single situation from becoming a human error mishap. Correcting the issue with Ownership has the potential to prevent dozens of human error mishaps in the future.

What should be noted in this typical example is that none of the actions that must be taken to correct this high-risk situation are the responsibility of the storekeepers. Reaffirming process ownership and clearly defining the necessary process to be taken on weekends and second shift are not storekeeper required actions. Even though the confusion within this process did not originate with the storekeepers, had a storekeeper dropped and damaged one of these critical pieces of mission equipment, a Mishap Review Board would have, without doubt, found the storekeeper at fault for not following the "Two-Person Lift" handling directions on the box. This example demonstrates exactly how performing a CoBRA Risk Assessment benefits the local workgroups. When workgroups identify process weaknesses, it allows the flawed process to be corrected before there is a human error incident. This protects workgroup members from potential blame, or even injury, should a mishap occur when working the flawed process.

18 - LITTLE THINGS CAN BITE HARD

A company needs to arrive at a point where local workgroups can perform risk assessments on their own processes without outside assistance. A local workgroup such as janitors/custodians needs to be able to sit down with their supervisor and discuss if there are any situations that the group could think of where a human error by one of the group members could hurt themselves, hurt others, or have a serious negative impact on the company. The group needs to discuss any incidents of "close calls" or "near misses" they encountered themselves or heard about from others. For example, this group discussion might lead to an issue such as one of the custodians pointing out that their cleaning carts contain bottles of both bleach and ammonia, and a new employee to the group almost added both bleach and ammonia together into the same bucket. Fortunately, the new employee was stopped in time and no harm was done. The new employee to the group was not aware that the mixing of both bleach and ammonia creates a toxic gas that could injure or kill.

After identifying a scenario with the potential for a serious situation as a result of human error, the workgroup might then look into what, if any, processes exists to assure that all new members of this workgroup understand the danger of mixing ammonia and bleach. The workgroup may determine that the only process they now have to prevent this from happening is just hoping that the Human Resources department would only hire experienced individuals that already knew of the bleach and ammonia danger. The workgroup would then do a CoBRA 30 Minute Risk Assessment for the process of just hoping a newly hired workgroup member would already know of the hazard based on Human Resources hiring criteria. After asking the questions in the twelve item checklist, the potential Consequence for determining the risk assessment value might be death/serious injury. From the management supplied list of Consequence values, the assigned value for death/serious injury is the maximum value of a 5. This workgroup would then look at the list of Control Strengths and the assigned matching Likelihood values. It would be obvious to the workgroup that a process with a control of merely hoping something is done, fails badly when running the twelve item checklist questions. The twelve item checklist questions would show this process to have extremely weak controls, and therefore would be assigned a Likelihood value of 5. The final risk value for this CoBRA 30 Minute Risk Assessment would be 5 X 5 = 25. Clearly this is a very high-risk situation and the workgroup needs to evaluate this situation further to see how the risk can somehow be lowered.

Searching for a way to lower the risk, the workgroup might then look at the company-provided list of Likelihood/Control Strength values and might decide that the best choice for their particular situation is a "Checklist with Employee Sign-Off", with the company supplied Likelihood/Control Strength list giving it a value of 2. A Consequence value of 5 and a Likelihood value of 2 would lower the final risk assessment value to a 5X2=10 and a significant improvement over 25. The workgroup of janitors/custodians might then agree that a process is needed to notify new employees of the hazard of mixing bleach and ammonia and they have determined that the group will create a new employee orientation class whose attendance is mandatory and verified by employee signature at completion. This orientation class will point out the danger of mixing ammonia and bleach, as well as any other potential dangers the group feels a new hire needs to be made aware of. The group might then run the CoBRA 30 Minute Risk Assessment again, including the new employee orientation class with signed buyoff as part of the process, and verify by using the twelve item checklist questions that the risk value has been truly lowered.

For a second possible situation, this same workgroup might also determine that there is a more expensive cleaner available that does not contain ammonia. With additional budget funding, the more expensive cleaner could be purchased and the risk eliminated completely. The CoBRA 30 Minute Risk Assessment could be provided to management as the rationale for requesting the additional funding based on lowering the risk further. The CoBRA 30 Minute Risk Assessment can serve as a tool to assist multiple levels of management in their decision making.

A company's management may wonder if its various workgroups will buy into doing risk assessments on their own processes. Management may have been burned in the past when they have tried other Quality/Safety/Reliability improvement methods, but were unsuccessful in getting the individual workgroups to make a serious effort to implement them. What is different about asking workgroups to do a CoBRA 30 Minute Risk Assessment is that it is not that difficult nor too time-consuming. It would also be quickly apparent to workgroups that a half-hearted effort that only finds a very limited number of processes with a potential for serious consequences, is actually telling upper management that what the workgroup does may not be very important to the company. It is clear to everyone that company resources tend to flow to workgroups that perform important functions and away from workgroups that do not. It is much more likely that workgroups will go to great effort to perform as many CoBRA 30 Minute Risk Assessments as possible in order to highlight to upper management any critical functions their particular workgroup performs for the company.

What is most important is that by doing CoBRA 30 Minute Risk Assessments, workgroups will have to look closely at the processes they perform and realize the potential worse case scenario if a human error occurs in the process. The *Process Owner* of each process will be established and it will be made clear who is capable of changing the process, and more importantly, who is not permitted to change the process. Workgroups will identify the pieces of the process that are vital to preventing human error and will realize that when those pieces are missing or weakened, the likelihood of human error dramatically increases. The workgroups will also identify any areas where they are vulnerable to the actions of others and realize that special care must be taken when dealing with outside people and processes over which their workgroup has no control.

Another reason CoBRA 30 Minute Risk Assessments work well with local workgroups when other types of Risk Assessment fail, is that a CoBRA 30 Minute Risk Assessment does not require the unrealistic action of having local group members "guess" on when they think the next human error and the serious consequence is going to occur. You can't get a usable and repeatable Likelihood value by asking a group of welders "When in the future do you guys believe you are going to make a bad weld that will injure or kill someone?" It is meaningless to ask this group if they think one of them is going to make a bad weld that could kill someone in the next 10 welds, next 1000 welds, or even within the next 10,000 welds. That is a question they don't have the capability to accurately answer, and how much confidence can you have with something that is just a guess?

A CoBRA 30 Minute Risk Assessment would not ask the local workgroup of welders to guess on when they are going to make a bad weld that will kill someone. A CoBRA 30 Minute Risk Assessment will instead ask questions like "Do you work to written procedures or verbal instructions?", "Does your group go through any type of training and is there a test at the end of the training?", "Is there any type of inspection done to each of your welds, and how knowledgeable are the guys doing the inspections?", "Are the welding machines that you use of good quality, and do these machines ever give you problems?". A group of welders cannot be expected to tell you when in the future they are going to make a human error that is going to kill someone, but they can answer simple basic questions about their daily jobs and that is all that a CoBRA 30 Minute Risk Assessment needs in order to give an accurate and repeatable risk assessment value.

A company should not plan on seeing a dramatic improvement in reducing the negative impacts from human error mistakes by just having one or two people doing investigations on incidents after they have occurred. A significant improvement will only happen when each workgroup within a company has carefully reviewed all of their own processes in order to assure that they are being worked exactly as the owner of that process intended.

Each workgroup needs to have the mentality that they are living in a world full of risk, and they alone are responsible for recognizing where and how human error is a threat to their own group members and to the company that employs them. Workgroups need to quit thinking that human error is something that sometimes happens to other people when they are careless, but will not happen to their workgroup because they are smarter and more careful than others. A workgroup cannot assume that because they have yet to have a serious human error mishap, that they are somehow immune to future problems.

19 - NO PROBLEMS, THINK AGAIN

By now, it should be apparent that there is a significant difference in the basic concept of the common types of risk assessments compared to the CoBRA 30 Minute Risk Assessment. The most obvious difference being that in a CoBRA 30 Minute Risk Assessment it makes no difference if the process has a history of no problems. Where most common risk assessments rely heavily on any kind of problem history, a CoBRA 30 Minute Risk Assessment only cares about how capable are the processes presently in place for dealing with the potential of human error.

You cannot have confidence in a process just because the process has no history of having problems. I like to sleep in a secure home and therefore every night before going to bed I lock my doors, turn on my outside lights, and activate my alarm system. For good measure, I also have a pit bull with a bad attitude that sleeps next to my bed. Let us assume that I have a next door neighbor that does not lock his doors, does not have outside lighting, does not have an alarm system, and does not have a dog. This neighbor has his own unique way of securing his home. Every night at sunset, this neighbor walks out onto his front yard wearing only his underwear and loudly shouts "All you crooks stay away".

My house has never been robbed, but neither has his. Does the fact that my neighbor has never been robbed, mean that standing in your yard wearing only your underwear while shouting "All you crooks stay away", is a suitable process for securing your home? If your definition of having a good process is not having any problems, then my neighbor's process for securing his house is every bit as good as mine. In fact, it could be argued that because my neighbor's process does not incur the cost of security monitoring, the cost of electricity for outside lighting, or the cost of feeding a dog, that my neighbor's process is actually superior to mine.

I feel confident that the only reason my neighbor has not been robbed is that criminals have never really attempted to rob his house, and therefore his unique security system has never been challenged and given the opportunity to fail. Not having problems can be the result of a well designed and properly implemented process, but it can also be the result of a process never having been sufficiently challenged. A history of no problems really doesn't mean that much at all. The only way to have true confidence in any process is by having it analyzed by experts. The best experts are the local workgroups that are tasked with actually working the process. A CoBRA 30 Minute Risk Assessment is a simple, quick, and accurate method, well suited for use by the local workgroups, in order to analyze their own processes.

It is a common practice to look at a history of no problems as justification for not worrying about a process. A referee in a boxing match will always give initial directions to the boxers at the start of the match during which he clearly stresses "Protect yourselves at all times". The concept of protecting yourself at all times should be the major guiding principle for every company when it comes to dealing with the potential of human error. The processes a company should be truly afraid of are those processes that are not being looked at because they have no history of having problems, and therefore considered low risk.

Workgroups have to be made to understand that the potential for human error always exists and it is up to the group itself to assure that their processes are all well designed and well implemented in order to deal with that potential. Performing risk assessments is the best way to address the potential for human error. A workgroup failing to acknowledge that there is always the potential for a human error will only leave the workgroup that much more vulnerable to the consequences of the human error.

The following is another example of the difference between a CoBRA 30 Minute Risk Assessment and other common risk assessments. You go to your doctor for a medical problem and get a prescription that you take to a pharmacist to have filled. Typically we just trust that the pharmacist correctly reads the doctors handwriting and also hope that the pharmacist does not make an error filling the prescription with the right medication and dosage.

A common risk assessment that relies on historical failure data would say that millions of prescriptions are filled daily without error, and therefore the chances of the pharmacist giving you the wrong drug, or the wrong dosage, is extremely rare. Based on this history of extremely rare instances of human error, the common risk assessment would have a minimal

Likelihood value and thus a low final risk assessment score for this process.

With a CoBRA 30 Minute Risk Assessment, the millions of prescriptions being filled daily without human error don't matter. What matters to a CoBRA 30 Minute Risk Assessment is only if you have capable processes in place that can adequately deal with the potential of the pharmacist making a human error in your particular situation. Processes being performed by others that you are just assuming will be performed properly have little or no value in reducing the Likelihood value in a CoBRA 30 Minute Risk Assessment. Even though this pharmacist may have internal procedures that check and double check for mistakes, CoBRAs 30 Minute Risk Assessments only place value on processes that are owned and controlled by you. In this case, since the consequence of a mistake could be death, the CoBRA 30 Minute Risk Assessment would have a final risk score that would be very high.

When a low risk assessment score is determined, a company usually requires no additional actions. When a high risk assessment final value is determined, some actions are typically required. In this particular example, only the CoBRA 30 Minute Risk Assessment would lead to requiring further actions. When looking at this situation, the obvious question is "Can anything be put in place to make me less vulnerable to a pharmacist making a human error mistake while filling my prescription?" After evaluation, a potential improvement might be to ask the doctor during the initial visit exactly what drug and dosage he was calling for when he gives you the prescription. By comparing what the pharmacist gives you to what the doctor said he was going to give you, would eliminate part of the risk in this situation. Making this change is a significant improvement towards lowering the risk, but it only removes the potential problem of the pharmacist incorrectly reading the doctor's handwritten instructions. There remains the possibility of the pharmacist knowing exactly what the doctor wanted, and still making a mistake by putting the wrong drug or dosage in the bottle. As a result of the still existing human error potential that is not being addressed, the subsequent CoBRA 30 Minute Risk Assessment final value would still be high. You are now left with the decision of either accepting the remaining high risk or refusing to take the medicine from the pharmacist.

It is pretty clear that nearly all of us would just accept the high risk. We would probably just live with the assumption that an error was not made and take the medication provided by the pharmacist. What is important here though is that we are now making a better-informed decision and we did significantly lower the initial high risk. The fact that we will now ask the doctor first about the prescription has significantly lowered the overall risk and that is far better than if we just called it a low-risk to start with and did nothing else.

History is not important with a CoBRA 30 Minute Risk Assessment. The fact that a pharmacist has not made a mistake in the previous 25 years of filling prescriptions does not mean that he will not make a mistake today. Humans are just too complicated to ever feel confident about future actions based on successful past performance. In the previous 25 years, this same pharmacist may never have had the things going on in his life that is happening now. Today this pharmacist may be dealing with a wife that is divorcing him, or a father with cancer, or a business that is failing due to financial problems. You never know what a person is currently dealing with that may cause his performance today to be different than what it had been in the previous 25 years when he never made a mistake. CoBRA 30 Minute Risk Assessments work because they accept the fact that humans are complicated and they are always capable of making an error.

Another everyday example that helps show the difference between a CoBRA 30 Minute Risk Assessment and other risk assessments: I need to fly from Orlando to New York. I know that it involves buying a ticket, boarding a plane, and then just sitting there trusting and hoping that the pilot avoids making a human error on this particular flight. A common risk assessment relying on historical failure data would point out that millions of passengers fly every day without problems, and the odds of me dying in a plane crash are extremely remote. Because of the low Likelihood value of dying in a plane crash based on the successful history of commercial flying, a typical common risk assessment would come back with a low final risk value.

The CoBRA 30 Minute Risk Assessment would ask what processes that I own and control are capable of dealing with the potential of this pilot making a human error. I intend to fly on a large U.S. airline, so I can assume that this airline and the FAA / NTSB have suitable processes in place to address pilot error, but these are not processes that I own and control. A typical risk assessment would determine a low risk based on airlines history of success, and therefore no need to be concerned about getting on this plane and flying to New York. A CoBRA 30 Minute Risk Assessment would determine a high final risk value based on the possibility of death from a potential pilot error and my lack of personal ownership and control of any processes capable of dealing with the potential for pilot error. The CoBRA 30 Minute Risk Assessment high risk value would warn me that I need to evaluate this situation further. As a result of the high risk assessment value, I would look more closely at my planned flight from Orlando to New York. The flight I had chosen involved making a stop in Atlanta. Since takeoffs and landings are the most dangerous part of flying, I realize that a direct flight to New York would have one less takeoff and landing. I could reduce my risk by choosing a direct flight that avoids an additional takeoff and landing. The direct flight is probably more expensive, so now I must make a decision if I consider the lower risk to be worth the increased cost of the flight.

The important differences in this example are that relying on historical failure data would not have raised any flags and there would have been no reason to evaluate the flying process any further. The CoBRA 30 Minute Risk Assessment would have dictated the need for more evaluation and this additional evaluation found that the risk in this process could be significantly lowered with minimal effort. Remember, the Likelihood value in a CoBRA risk assessment is based on there being any problems with Process Ownership, Process Design (DATOM), or Process Control Strength. If you are in a high Consequence situation and find yourself dependent on a process where you have no knowledge as to if there are problems with Process Ownership, Process Design (DATOM), or Process Control Strength, then you should be very concerned. You are never going to attempt to reduce risk if you fail to realize when a high-risk situation even exists.

It is important that people begin to understand that it is a high-risk situation anytime the consequences are serious and you find yourself depending on other people and processes that you have no control over. When the consequences are serious, people need to be less trusting and less comfortable when it comes to depending on others.

20 - ACCEPTING HIGH-RISK SITUATIONS

The discovery of a high-risk situation during a risk assessment should be treated by management as a fortunate opportunity, rather than something negative. Whenever a potentially serious situation that could result from a human error within a process is identified, the question becomes to either accept the risk as is or change the process to lower the risk. Qualified people determining which high-risk situations to accept and which to correct should be a part of the everyday normal business for a company. No business could ever be commercially viable if they are not willing to live with many high-risk situations. Every organization has to work with a finite amount of resources and having the right people determining which high-risk situations to accept and which high-risk situations to correct can easily determine which companies survive and which ones fail.

Mishap investigation boards always find fixes that, had those fixes been in place originally, would have prevented the serious incident now being investigated. Sadly, far too often these fixes are simple and of very little cost to implement. So why is it that serious incidents occur in a process as a result of human error when there are simple fixes available that could have been in place originally and prevented the incidents?

Risk assessments are the best tool for identifying and preventing potentially serious consequence situations, but only if they are properly performed. An organization is doing something definitely wrong if a serious consequence has occurred that could have been prevented with a simple, low-cost fix. These organizations need to step back and re-think how they are doing their risk assessments. When risk assessments are being done right, if there were any simple low-cost fixes possible they would have been identified and already put in place.

Organizations need to be honest with themselves and admit that Likelihood values based on the "odds" of something occurring are not much more than guesses and not any kind of reliable hard science that should be relied upon. In addition, you cannot expect honest results from individuals performing the risk assessments when their management is going to be pleased with them when they come up with a low Likelihood value and going to be upset with them when a high Likelihood is assigned. It is very easy to come up with the rationale to justify arriving at a low Likelihood value, especially when it pleases their management. It is not fair to individuals performing risk assessments to put them in that situation. This is often why mishap investigation review boards looking into serious incidents find that simple, low-cost fixes were available that would have prevented the incident but were never even considered. Low Likelihood risk assessment values can prevent the need or incentive for the identification and implementation of even simple, low-cost fixes to flawed processes. The purpose of doing risk assessments should not be to provide a means of justifying not taking any actions at all.

When using the CoBRA process, the responsibilities

of the individuals involved is clear. The person performing the CoBRA assessment identifies a process where a human error during the process can result in a serious consequence and then:
- Evaluates the process that is in place intended to prevent any human error to determine any issues with Process Ownership, Process Design (DATOM) or Process Control Strength.
- If no issues found, the process is identified to be low-risk. No further actions required.
- If issues are found, but they can be corrected locally, then correct those issues and identify the process as low-risk. No further actions required.
- If issues are found and they cannot be corrected locally, identify the process as being high-risk and forward the assessment to an individual with the authority to accept high-risk situations.

Individuals with authority to accept high-risk situations will review CoBRA assessment information and then:
- Decide if the high-risk situation warrants additional actions and if so, provide direction and resources to assure those actions are accomplished. Identify as low-risk when all actions needed to correct issues have been completed.
- If it is determined that additional actions are not feasible or warranted in their judgment, then assume responsibility and accept situation "as is". Document rationale for requiring no additional actions. Identify as an accepted high-risk situation.

Anytime there is a potential for a serious consequence as a result of human error combined with a process weakness involving Process Ownership, Process (DATOM), or Process Control Strength, then this situation must be corrected or elevated. Even if the individual performing the risk assessment believes that the weakness in the process is extremely unlikely to actually result in the serious consequence, it makes no difference in a CoBRA risk assessment. When there is a potential for a serious consequence should a human error occur, and the process whose function is to prevent the human error is flawed or weak, then a high-risk situation exists.

Only the individual with the authority to accept a high-risk situation is allowed to make a final decision on a process with a potential serious consequence combined with any type of problem or weakness regarding Process Ownership, Process Design (DATOM), or Process Control Strength. The individual with the ability to accept risk can justify their decision using any criteria they wish to use, but that does not change the level of risk. An accepted high-risk situation does not now become a low-risk situation. If the flaws and weaknesses in the process have not been corrected, only accepted, the high-risk situation still exists.

There is nothing wrong with a decision to accept a high-risk situation as long as it has been made based on a thorough evaluation and the determination has been made that there is nothing reasonable that can be done additionally to reduce the risk. It is the unidentified high-risk situation, one that has never been closely evaluated, that is the real danger to any organization. Good management will have very qualified and capable individuals who understand and accept this responsibility of deciding either to live with or correct high-risk situations. When an individual understands that they alone are responsible for accepting a high-risk situation, they are far more likely to make sure that there truly are no other reasonable choices available in this situation. You can be assured that these individuals tasked with accepting high-risk situations will require that any simple and low-cost preventable actions will always be implemented when available rather than having to just accept a potentially serious situation.

The CoBRA risk assessment process will result in the identification of many high-risk situations that will be eventually accepted after additional evaluation by individuals trusted to make those decisions. The fact that these potentially serious situations have been accepted, does not remove the original potential for a serious consequence. The only difference now is that a decision has been made that the organization will live with this situation and not correct it. The typical reason for an organization to accept living with a potentially serious situation is that a determination has been made that there is no simple, low-cost fix to this situation and the situation does not warrant a difficult, high-cost fix. Again, there is absolutely nothing wrong with this process and that is just how business must work in order to survive.

These evaluation records of accepted high-risk situations should be maintained in a database in a manner to facilitate easy review in the future. These records should be treated the way a police department maintains "Cold Case" records on crimes they could not solve with the technology and personnel available today. Many old crimes were finally solved when improvements in DNA technology came along and provided additional information to many cases. Similarly, technology may improve in a manner that someday will allow for a simple fix to a potentially serious consequence situation that today is too difficult or expensive to correct. Every record in the database of accepted high-risk situations should be periodically re-evaluated to see if any new advances in technology can allow for the situation to be corrected rather than just accepted.

21 - TO SUM THINGS UP

The following is a quick summary of the most important things our group of engineers learned over 24 years of doing thousands of human error investigations:

Don't make dealing with human error any more complicated than it has to be – There are only a few basic parts of any process that you really need to focus on: Process Ownership, Process Design(DATOM), and Process Control Strength. Remember, the *Process Owner* has some goal he is trying to achieve and has designed a plan requiring specific actions he wants the *Process Workers* to take in order to achieve the goal. The *Process Owner* has chosen some type of controls that he feels are appropriate for assuring that the *Process Workers* are made aware of his plan of desired actions in order to implement the actions exactly as desired.

A high-risk situation is anytime there is a potential for a serious consequence if a human error failure occurs and the processes that must function correctly in order to avoid the human error are poorly enforced, poorly designed, or poorly implemented. Poorly enforced means a process fails Process Ownership review. Poorly designed means a process fails Process Design (DATOM) review. Poorly implemented in a high-risk process means the controls being used in the process are flawed or not strong enough to match the seriousness of a potential human error failure occurring when working the process.

Processes that are **unlikely to fail** due to human error must have all of the following:
- No problems with Ownership
  - *Process Workers* understand that process directions are hard requirements that can only be changed at the Process Owner's direction.
- Properly Designed
  - Correct and complete (DATOM) Define, Assign, Train, Organize, and Monitor.
- Adequate Control Strength
  - Strong controls are being used to convey and implement the *Process Owner's* desired actions.

Processes that are **likely to fail** due to human error can have any of the following:
- Ownership Problems
  - *Process Workers* believe process directions are open for "modification" by the *Process Workers* rather than hard requirements that must be exactly complied with.
- Design Problems
  - Poorly done Define, Assign, Train, Organize, or Monitor.
- Control Strength Problems
  - Weak controls being used to convey and implement the *Process Owner's* desired actions.

A CoBRA 30 Minute Risk Assessment quickly identifies if any of these conditions exist. Once a company has provided all of its workgroups with both a 1-5 Likelihood ranking of the company's controls based on strength and a 1-5 Consequence ranking based on the company's unique definition of what conditions are considered serious to the company, a final risk value can be determined.

Remember, the overall goal is preventing the serious consequence from a human error in a process and not necessarily the avoidance of all human error itself. Often it is far more affordable to accept the possibility of human error at one point in a process as long as it does not occur in other parts of the process.

You can avoid the serious consequence by having a process that either prevents the initial human error or by identifying that a human error has occurred in a timely enough manner that allows for the averting of the potential serious consequence. For example, it might make more business sense to put the money and resources into the inspection and testing of the final product, rather than making sure that there is never a human error made at each and every step in the production of the product. You can often use lower cost weaker controls early in a process as long as you have very strong controls at the end of the process.

If a bad product is ever found during final testing you still need an investigation as to why the error occurred and correct that problem, but everyone should keep in mind that the overall process was designed by the

Process Owner to prevent a bad product from getting to the customer, rather than preventing any bad product from ever being made. You do not have to always prevent every human error in order to have a well designed and suitable process, but you do always have to prevent the potential serious consequence of the human error.

You cannot have true confidence in any major critical process if you do not first have equal confidence in all of the supporting processes required to accomplish the major process. It is often the failures and weaknesses in the supporting processes that result in problems with the major critical processes. It must be clearly understood how processes are all dependent on each other. If a local workgroup is failing to address problems in its own processes, it might also be creating a potential failure in processes belonging to other groups. One workgroup cannot have complete confidence in its own processes if those processes are dependent on other processes being provided by another workgroup. If you don't have control over a process, you are then dependent on those who do have control over that process. Real confidence only comes from having risk assessments performed on all processes that have the potential to result in a serious consequence if a human error occurs during the process.

Unidentified high risk in a process is similar to a person not realizing that they have high blood pressure. The person may still be feeling fine, but there is a serious situation going on with the high blood pressure that can eventually result in deadly consequences. A process with unidentified high risk, just like a person with unidentified high blood pressure, both may appear to be doing fine with no problems being presently observed, but all the while a potential for serious consequence exists. People often do not see a need to go to a doctor when they are feeling fine. In a similar manner, workgroups do not typically feel a need to analyze and modify processes that are not presently having problems.

It was not that long ago that the only way a person could determine if they had high blood pressure was with an instrument that measured blood pressure using the height of a column of mercury. This device was expensive, complicated, and it took a specially trained doctor or nurse in order to get an accurate reading. Typical risk assessments are like those early blood pressure measuring devices in the sense that they are also costly, complicated, and require specially trained individuals to get usable results.

Today, people are far more likely to be aware when they have high blood pressure. What changed over the years was that more people became informed that feeling good meant nothing when it came to having high blood pressure and therefore they realized that they needed to have their blood pressure periodically checked, even when they were feeling fine. Another significant change was the development of inexpensive and simple devices for checking blood pressure that have now become available to everyone to regularly use at home. No longer did it require a doctor or nurse, using an expensive and complicated instrument, to determine if a person had high blood pressure. A CoBRA 30 Minute Risk Assessment can do the same thing for identifying high-risk processes within a company that the development of inexpensive and simple blood pressure monitoring devices did for identifying people at risk due to high blood pressure.

A CoBRA 30 Minute Risk Assessment allows local workgroups to monitor their own processes without the need for highly trained outside experts. The members of local workgroups are the true experts on their own specific processes, and having a detailed understanding of the process being analyzed is 99% of any risk assessment. Local workgroups already have the knowledge to accurately answer any questions regarding their processes. What these local workgroup experts presently lack is the understanding of what types of questions they need to be asking themselves in order to identify processes that are weak and flawed. A CoBRA 30 Minute Risk Assessment leads these workgroups into asking themselves the right questions.

Whenever there is a mishap investigation, failed or weak processes will be uncovered identifying contributing causes to the mishap. Every organization has some flawed processes, so the goal becomes finding and correcting them before they lead to a serious consequence. An organization needs to look at what methods they are presently using to discover flawed processes instead of having a mishap investigation tell them later after a serious incident has already occurred. Waiting for mishap investigations to discover flawed processes, is not a pathway to success. It should be everyone's goal to identify weak and flawed processes using risk assessments, rather than eventually finding these same weak or flawed processes later as a contributing cause in a mishap investigation.

Strong controls absolutely have to work every single time. Whenever there is an instance when a strong control fails and there is not a serious incident, that should be considered a lucky gift your organization has been given and the opportunity should never be wasted. If a strong control fails, and it does not result in a serious consequence, that does not lessen the seriousness of the strong control failure in the slightest. Strong controls are what keep people from dying or companies from experiencing a devastating impact, so you have to completely understand the cause of each and every strong control failure. Serious actions are required to be taken immediately to prevent a recurrence. Strong control failures cannot be allowed to occur for any reason.

People who have ridden motorcycles for many years have the kind of mentality that is necessary to be a good *Process Owner*. Experienced motorcycle riders know that when you start to believe that other drivers on the road will not make a mistake, is when you will find yourself flying in the air over the hood of a car. Experienced motorcycle riders understand that following the rules and personally doing everything right, is still not enough to survive. These older motorcycle riders know that they themselves cannot afford to make a mistake, and at the same time, they also have to ride in a manner that protects them from the mistakes of other drivers. That is the mentality that a good *Process Owner* is required to have when there is a serious consequence potential. A good *Process Owner*, for a serious consequence process, grasps that it is not enough to just build a process that is capable of doing everything right. A good *Process Owner* also understands that they have to build a process that can deal with those working the process doing something wrong.

Human error should be thought of as merely a disconnect between the actions the *Process Owner* expected and the actions that were actually taken by the *Process Workers.* Processes are going on continually and in some processes, you may be the *Process Owner* and in other processes, you may find yourself as the *Process Worker*. The potential for disconnects always exists. Every person, in every organization, from the CEO down to the window cleaners should be wearing a T-Shirt that says **Warning - I am a human and I forget things, I get confused, and I get distracted, so just deal with it!** You should never look at a human error incident as "Why did this person screw up?" but rather as "Why did this process not happen exactly as the *Process Owner* desired?" When our group first formed, we originally looked at investigations like some crime had been committed and our job was to gather evidence to help "convict" the responsible individual who failed to follow some often obscure requirement that may have existed somewhere. This was so very wrong, and it only resulted in making it more difficult to work with these same people in the future. It is much better to think of any incident as a situation where everyone involved truly wanted the process to work perfectly, but somehow it didn't, and now we just have to determine how to make sure it will always work well in the future.

Changes are always taking place in any organization. Some changes do not affect risk, but many changes do. The local workgroups must be able to recognize changes that have the potential to negatively affect risk. It is the local workgroups that must be able to recognize that a change has occurred within the process that could affect risk and realize when there is a need to stop and reevaluate the process. With so many processes needing constant analysis, only the individual local workgroups are in the position to be able to keep watch over their own processes.

You cannot have confidence in a process just because the process has no history of any problems. *Process Owners* of high consequence situations should never be able to avoid addressing the potential for human error by just saying that it is very unlikely a human error will occur because there is no history of that error previously happening. A past history of no problems is not a "Get Out of Jail Free Card" that can be played whenever it is difficult to properly deal with a flawed process that also has the potential for a serious consequence.

A railroad company would never assume that a bridge was safe to use based solely on the fact that there is no previous history of it collapsing and sending a train into the river. A railroad company does not wait for a bridge to fail, they perform periodic inspections and they have specific criteria to which they inspect to. The results of the bridge inspections determine if, and in what order, the railroad company will do repairs on the bridges. Companies need to look at all of their processes like they are bridges supporting other processes.

All processes need to be periodically inspected to specific criteria to assure the processes are working as desired and will not fail and negatively impact themselves or the other processes they support. A CoBRA 30 Minute Risk Assessment serves as a form of process inspection tool. As a process inspection tool, it has specific inspection criteria which are the twelve item checklist questions that assure there are no problems with regard to Process Ownership, Process Design (DATOM) or Process Control Strength. Like the bridge inspections, the CoBRA 30 Minute Risk Assessment provides a means, in the form of the final risk value, that allows management to determine if, and in what order, repairs to processes are to be performed. When there is a potential for serious consequences, a company cannot afford to continue running processes until they fail and then do mishap investigations to determine why they failed. You cannot trust any process that has not been analyzed and determined by that analysis to be functioning without flaws.

An organization that wants to prevent serious consequences resulting from human error needs to do the following:

- Identify which processes have the potential to

result in a serious consequence if a human error occurs.
- Evaluate those processes through the COBRA risk assessment procedure to assure there are no problems with Process Ownership, Process Design (DATOM), or Process Control Strength.
- If any issues are found, correct those issues. If they cannot be corrected at the working group level, then elevate as a high risk to individuals who have been given the authority to accept high-risk situations.
- Allow only clearly designated and highly qualified individuals determine if a high-risk situation due to a problem or weakness with Process Ownership, Process Design (DATOM), or Process Control Strength can be corrected or requires acceptance and is allowed to exist as is.
- Maintain a database of all accepted high-risk situations and review it regularly.

I am going to end this book here, but before I do I want to thank each and every person I have ever dealt with during my 24 years on the Space Shuttle Program. I want to say to the other members that were in my group, to my management, to my NASA counterparts, and to the technicians and engineers that you all were the finest group of people that anyone could ever dream of working with. The entire Shuttle workforce was professional to deal with, but at the same time, it was like being around good friends. This group of people made it a pleasure to get up every morning and go to work. When the Shuttle Program ended I honestly felt like I was saying goodbye to members of my family rather than to coworkers. It is a shame that everyone can not experience how truly wonderful a job can actually be when you get to work with people like I had the pleasure of working with on the Shuttle Program.

www.ingramcontent.com/pod-product-compliance
Lightning Source LLC
Chambersburg PA
CBHW020913180526
45163CB00007B/2712